GENDER INTEGRATION IN
NATO MILITARY FORCES

Gender in a Global/Local World

Series Editors: Jane Parpart, Pauline Gardiner Barber
and Marianne H. Marchand

Gender in a Global/Local World critically explores the uneven and often contradictory ways in which global processes and local identities come together. Much has been and is being written about globalization and responses to it but rarely from a critical, historical, gendered perspective. Yet, these processes are profoundly gendered albeit in different ways in particular contexts and times. The changes in social, cultural, economic and political institutions and practices alter the conditions under which women and men make and remake their lives. New spaces have been created – economic, political, social – and previously silent voices are being heard. North-South dichotomies are being undermined as increasing numbers of people and communities are exposed to international processes through migration, travel, and communication, even as marginalization and poverty intensify for many in all parts of the world. The series features monographs and collections which explore the tensions in a "global/local world," and includes contributions from all disciplines in recognition that no single approach can capture these complex processes.

Previous titles are listed at the back of the book

Gender Integration in NATO Military Forces
Cross-national Analysis

LANA OBRADOVIC
University of Nebraska Omaha, USA

Routledge
Taylor & Francis Group

LONDON AND NEW YORK

First published 2014 by Ashgate Publishing

2 Park Square, Milton Park, Abingdon, Oxon OX14 4RN
711 Third Avenue, New York, NY 10017, USA

Routledge is an imprint of the Taylor & Francis Group, an informa business

First issued in paperback 2016

British Library Cataloguing in Publication Data
A catalogue record for this book is available from the British Library

The Library of Congress has cataloged the printed edition as follows:
Obradovic, Lana.
    Gender integration in NATO military forces : cross-national analysis / by Lana Obradovic.
        pages cm. -- (Gender in a global/local world)
    Includes bibliographical references and index.
    ISBN 978-1-4094-6476-1 (hardback)
1. North Atlantic Treaty Organization--Armed Forces--Women.
2. Europe, Western--Armed Forces--Women. 3. United States--
Armed Forces Women. I. Title.
    UB419.E8O47 2014
    355.0082'091821--dc23

                                                        2013044879

ISBN 978-1-4094-6476-1 (hbk)
ISBN 978-1-138-25165-6 (pbk)

Contents

List of Figures and Tables

Figures

Tables

Preface

Gender Integration in NATO Military Forces develops the discussions of gender, conflict and the military that thread through many of the books in the Ashgate series on Gender in a Global/Local World. Lana Obradovic takes a broad, comparative approach to understanding the integration of women in the various NATO military forces, including NATO's new member states in Eastern Europe. Her approach provides an opportunity to think beyond national borders, and to consider national, regional and international factors influencing decisions about women's participation and roles within the NATO military forces. The book seeks to understand the degree to which different democratic member states have integrated women into their military services, and the reasons behind these differences. It explores the various structural, institutional, cultural and international factors influencing states' military personnel policies on gender in NATO militaries. By examining the policies of all 24 member states, Obradovic develops an explanation for why particular states have (or have not) decided to integrate women in the military in ways that make their participation an integral part of the NATO military forces.

Gender Integration in NATO Military Forces challenges mainstream security studies' neglect of women and gendered perspectives and contributes to the growing field of feminist security studies. The book's analysis and argument are developed around four explanatory categories. On the state level, Obradovic examines the role of military manpower, domestic political and economic factors, and cultural factors. On the broader system level, she examines the impact of international factors on state policies and practices concerned with the integration of women soldiers into NATO military forces. Obradovic's systematic, comparative research demonstrates that civilian policymakers and military leaders in NATO are less affected by local gendered assumptions and practices and much more influenced by the need to meet recruitment numbers required for modernized and professionalized militaries, the demands of domestic women's movements, and the need to meet states' responsibilities under international agreements regarding gender equality and gender mainstreaming in the military. Obradovic concludes that masculine myths of soldiering are buckling under the weight of a new type of security environment that requires technical mastery and professional skills, and that military policies on gender integration in the military are more influenced by transnational women's movements' demands than by existing (local) gendered practices (and regimes). *Gender Integration in NATO Military Forces* is the first systematic gender analysis of women's integration into NATO forces. It raises important questions for mainstream, critical and feminist security studies, demonstrates the crucial role of comparative analysis and will no doubt inspire future research on the subject. We look forward to these debates.

Jane Parpart
Marianne H. Marchand
Pauline Gardiner Barber

Acknowledgements

I grew up listening to stories of my grandmother Mira's heroic exploits performing intelligence for Marshall Tito's partisans during World War II, reading books immortalizing the female protagonists of Yugoslav national liberation, singing lyrics such as "*mlada partizanka bombe bacala*" ("*a young Partisan woman threw grenades at the enemy*"), and shooting guns from the age of 9. In April 1992, as a new war broke out in my homeland, there was a knock on the door of my family's apartment in Mostar, Bosnia and Herzegovina. A group of uniformed men from Bosnia's Territorial Defense Forces, desperately seeking potential military recruits, unfolded a piece of paper with my name on it. They knew only that I was a sharpshooter who had trained three times a week for five years at the local range. I was 14—too young for legal recruitment—so they left, no further questions asked. They did not consider recruiting my father, Ahmet, a towering figure in both physical and intellectual dimensions, as he had no military and firearms training. My upbringing and this episode shaped my understanding of military recruitment in terms of manpower needs and largely devoid of any thoughts on societal and cultural standards of what were gender-appropriate roles and behaviors. We all fight as citizens of a state, regardless of our gender, I thought. Years after my own escape from the war in Bosnia, as a PhD student seeking to specialize in international relations and security studies, I joined Dr. Joyce Gelb's class graduate course on gender policymaking at the Graduate Center of City University of New York in 2005. I was not especially interested in gender(ed) variables, but as I set out contemplating my term paper topic on women and security, I began to rethink not only my own personal views but the way my field approached the analysis. Just as my thoughts evolved, so has this book over many years and I have accumulated a few debts that need to be acknowledged. I must thank Dr. Gelb, who along with my grade on this short paper on women's movements and military integration, wrote that one day it could make a good dissertation topic. She was the first one to suggest I pursue this interest. It was not until my other topic failed to excite my advisers that I approached Dr. Irving L. Markovitz—Lenny—with the question of gender integration in the military. I am greatly indebted to him as my adviser, mentor and a friend for his incredible support, advice and guidance. His encouragement and belief in me and the value of this project helped me complete that dissertation first, and for years after that, kept me rethinking my original analysis as he continued to challenge me to refine my arguments, expand the scope, clarify the logic of my method and inquiry, and take ownership of my findings. Without Lenny's insistence and conviction that I ought to continue my research on the subject, by writing papers and a book—this project, completed between my lectures on two continents—would not have materialized. I am grateful to Dr. Kenneth Erickson for teaching me how to edit, for his review and suggestions along the way, and Dr. Peter Liberman for his invaluable comments on research design, methods and analysis. I owe gratitude to NATO delegates for their willingness to share their data with me, and taking time to talk to me about the gender integration process in their states.

I would like to express my gratitude to the many people who saw me through this book, including Insung Lee, Chancellor of the Wonju Campus at Yonsei University, South Korea, my colleagues at the East Asia International College and Mercy College in New York for providing intellectual and institutional support. Without conference funds provided by these institutions, many of my presentations of this work at the Midwest Political Science Association and the American Political Science Association would not have been possible, and I would not have received invaluable suggestions and comments from Richard Eichenberg, Joshua Goldstein, Ann Tickner, Nikki Detraz, and Frank LeVeness. I would also like to thank Christina Wolbrecht, Amy G. Mazur and Kathryn L. Pearson, who inspired me to continue my exploration of the subject and writing when they bestowed on me the Sophonisba Breckinridge Award for the best paper delivered on women and politics at the 2010 Midwest Political Science Association Annual Conference.

Jane Parpart worked with me as Ashgate Gender in Global/Local World series editor, guided me through the process and helped me tease out my theoretical arguments as the final version of this book took shape. Her input on my drafts was instrumental in shaping and refining my ideas. I am thankful to Ashgate Publishers, editors and staff who have patiently waited for my drafts and encouraged me to finish. Moreover, I am indebted to an anonymous reviewer who shredded my first draft, and offered crucial comments that allowed me to pull it all together. There are a number of my colleagues who have made my days and nights writing this book less solitary, but I need to particularly thank my friend Billie Jo Hernandez for making it more fun.

I am grateful most of all to my family, to whom I dedicate this book. Thanks to my mom, Majda, the strongest woman I have ever known, who took me out of the jaws of war in Bosnia, and brought me to the United States. She gave me love and encouragement to pursue my goals, inspired and motivated me to be the best person, daughter, researcher and teacher I can be. I will always aspire to be like her. Thanks to my stepfather John for being proud of me the way my father would have been had he lived to see me succeed. And my sister, Lea, for making our lives more colorful and exciting, and for always reminding me to finish the book so that I can enjoy life.

I would like to thank my husband, David Candler, the greatest partner and supporter throughout this project, as most of the work occurred on weekends, nights, while on vacation, and other times inconvenient to my family. Thanks for continuously reminding me to write something readable and accessible, for reading and editing my drafts, listening to my incoherent research dilemmas and ideas, and keeping calm. Thank you for loving me and for always standing by me. And last, but certainly not least, to my sunshine, my daughter Liv, for reminding me every day that life is beautiful, and for understanding when Mama was writing this book instead of playing games. I hope that one day you will read this book and understand that you are my reason for it all.

Abbreviations

AK	Armia Krajowa
ANANDOS	National Association of Women Soldier Aspirants
ATA	Air Transport Auxiliary
AVF	All-Volunteer Force
CIF	Centro Italiano Femminile
CWINF	Committee on Women in the NATO Forces
DACOWITS	Defense Advisory Committee on Women in the Services
DCPC	Direct Combat Probability Coding
DoD	Department of Defense
DYSK	Dywersja i Sabotaż Kobiet
ERA	Equal Rights Amendment
FORSCOM	U.S. Army Forces Command
GDP	Gross Domestic Product
HDI	Human Development Index
IGI	Index of Gender Inclusiveness
IMS	International Military Staff
MC	Military Committee
MCWR	Marine Corps Women's Reserve
NATO	North Atlantic Treaty Organization
NCGP	NATO Committee on Gender Perspectives
NOW	National Organization for Women
NWLC	National Women's Law Center
OPWK	Organizacja Przysposobienia Wojskowego Kobiet
OWINF	Office on Women in the NATO Forces
PfP	Partnership for Peace
PSK	Pomocnicza Sluzba Kobiet
SPAR	Coast Guard Women's Reserve
SWAN	Service Women's Action Network
UDI	Unione Donna Italiana
UNDP	United Nation's Development Program
WAAC	Women's Army Auxiliary Corps
WAC	Women's Army Corps
WAFS	Women's Auxiliary Ferrying Squadron
WASP	Women's Airforce Service Pilots
WAVES	Women Accepted for Volunteer Emergency Service
WCTF	Women in Combat Task Force
WEAL	Women's Equity Action League

WFTD   Women's Flying Training Detachment
WMA   Women Military Aviators
WOPA   Women's Officers Professional Association
WREI   Women's Research and Education Institute

Chapter 1

Introduction

The military has traditionally been an all-male dominion. While there are many historical examples of women such as Joan of Arc, Margaret Corbin and an all-female Russian 588th Night Bomber Regiment fighting wars and standing beside men, for centuries men made up the majority of recruits and held the highest ranks. Most often, women joined or were recruited during wartime to help fill multifarious positions, from secretaries and nurses to bomber pilots and snipers. Once a war was over, however, the women were sent home, with their accomplishments relegated to historical archives. Popular literature most often discusses women's presence in the military and war in "sexier" roles as spies, saboteurs and nurses, while their combat experience has been airbrushed out of our collective memory.

Following World War II and during the early days of the Cold War, women were incorporated into the military services in most states on both sides of the Iron Curtain, but the presence, role and impact of these women were rather insignificant. In the majority of states, women performed their duties in administrative, personnel and medical positions. They were prohibited from attending military academies and were not allowed to scale the ranks to command levels. Most states had strict laws and regulations limiting the number of women soldiers and professional positions open to them.

However, since the mid-1970s and early 1980s this trend that had persisted for most of humankind's history has started to change worldwide. Numerous states have passed legislation permanently integrating women into their military forces and have dramatically increased women's numbers and the extent of their service. The number of female soldiers in NATO forces increased from 30,000 in 1961 to approximately 300,000 today.[1] Operations Desert Shield and Desert Storm in 1990–91 included more than 40,000 U.S. women soldiers and since 2003, 160,500 American female soldiers have served in Iraq, Afghanistan and the Middle East, meaning one in seven soldiers is a woman.[2] In 1970 the number of women in the U.S. military was a meager 1.4 percent, but by 1980 it had increased sixfold to 8.3 percent, and today women make up 15 percent of U.S. forces. Similarly, Spain increased its number of women in the military from 5.8 percent in 2001 to 13.47 percent in 2006, while Portugal's numbers rose from 6.6 percent to 12 percent. The change is particularly evident in NATO's new member states in Eastern Europe, such as Latvia, which boasts 17 percent, Slovenia 15 percent, and Hungary 20

---

1   Nielsen (2001: 30–34).
2   Solaro (2006: 15).

percent.[3] In fact, all NATO member states, with the exception of Iceland, since it does not have a standing army, have passed legislation permanently integrating women into their military ranks, and assigning functions and duties traditionally performed by men.

Not only are the numbers increasing but the roles that women play in the armed forces are expanding. The impact of these changes is best illustrated by former U.S. Joint Chiefs of Staff Chairman J. Vessey's observation that "the influx of women has brought greater change to the U.S. military than the introduction of nuclear weapons" (Carrol Hall 1993). Since the passage of the United Nations Security Council Resolution 1325 in October 2000, and 1820 in June 2008, states are legally bound to both recognize the importance of the role played by women in conflicts and in peace-building, and to increase women's numbers in peacekeeping, conflict prevention and resolution, prevention of sexual violence, as well as enforce gender equality on all levels of decision-making. Even though the NATO Committee on Gender Perspectives (NCGP) was established in 1976 and has provided advice and information to member states regarding employment of servicewomen, in 2009 its mandate was expanded to include supporting the implementation of Security Council Resolutions 1325 and 1820. The NATO multinational force deployed women for the first time in peace-enforcing and peacekeeping missions in Bosnia, Croatia and Kosovo. In Iraq, United States and British uniformed female soldiers served as members of search teams, while their Canadian counterparts continue to patrol check points in Afghanistan. The alliance continues to support gender initiatives through a variety of conferences and reports on integrating gender dimensions into NATO operations, and a recent report shows that today "NATO nations are therefore strongly encouraged to employ female personnel within the full spectrum of their operations"[4] (NCGP 2007: 11). Whether it is peacekeeping and peace-enforcing operations or new wars and missions fought in Iraq and Afghanistan, NATO member states have been seeking to reform their military services to integrate gender perspectives by passing policies that increase the quality of the experience for women in the armed forces.

Yet despite changes and initiatives on both domestic and international levels to integrate gender perspectives into the military, not all states have improved women's status in their armed services to the same extent. Some have erased all legal obstacles to women's progress in the military and have successfully promoted gender integration in the ranks, but others continue to impede it by setting limitations on women's equal access to careers, combat, ranks, or by simply

---

3 All of the data is available from the Committee on Women in the NATO website (now NATO Committee on Gender Perspectives), which publishes the annual report for each member state. Available at: http://www.nato.int/issues/women_nato/index.html Accessed August 14, 2008.

4 The same report points out that, although civilian women are often deployed as members of NATO and national personnel, the term "female personnel" is principally taken to refer to female members of NATO Nations' Armed Forces.

not seeking to change basic social and family friendly policies. Despite awarding a number of women medals of honor for their role in combat situations in Iraq and Afghanistan, the United States Department of Defense continues to ban women from engaging in ground combat. While Italian women were integrated in 2000, today they occupy only 3.9 percent of positions in the *Forze Armate*. Although women have served in the Turkish military since the time of Ataturk, the state today allows women only to enter as officers or non-commissioned officers, and most serve in administrative and medical areas. Greece established its Ministry of Defense Gender Equality Office only in 2004 to collect data and seek to eliminate gender discrimination. Some, such as Denmark, France and Belgium, still lack gender advisers, while others, such as Romania, have no programs or policies for the recruitment and retention of women. Among the newest members of the alliance, Poland and Romania show the least commitment, with just 2.1 percent and 4.6 percent of women in the military, respectively.

This transformation of government policies regarding gender integration in the military, no matter how imperfect, inadequate and far from complete, shows that the end of the Cold War ushered a new era of civil–military relations in democratic North American and European states. Governments are undergoing a rather difficult process of finding a new and suitable role for their military institutions in order to improve their effectiveness, strength and values, while simultaneously seeking to reassess their interaction with the democratic society they claim to represent and protect, and results vary from one state to another.

What explains the degree to which different democratic NATO member states have integrated gender into their military services? Why are women increasingly being integrated into some states' armed forces and not in others, despite an enormous organizational transformation, adaptation of facilities, change of physical prowess standards and interaction regulations—all very different from the integration into civilian employment where women have participated for many decades? What structural, institutional, cultural or international factors explain states' military personnel policy on gender, and which factors are the most important?

Such a significant variation in gender integration and women's participation in the military demands an explanation and much greater scrutiny. The last three decades have witnessed myriad social scientists and social commentators highlighting the reality of women in the military. It is not surprising that an intense public debate resulted in a significant increase in literature on the issue of gender integration in the military around the world. But what is surprising is that there are serious absences of empirical studies and comprehensive theoretical and comparative analysis in political science. It appears the state's gender integration policy in the military has fallen through the cracks between our levels of analysis and our increasingly narrowed subfields. By examining 24 NATO member states, the aim of this book is to build a theoretical model that explains why states abandon their policies of exclusion, and promote gender integration in a way that women's military participation becomes an integral part of military force.

It does so by drawing on both literature in international relations, in particular feminist security literature that questions systemic and gendered challenges of integration of women in highly masculine military institutions, and comparative gender policymaking literature that focuses on cross-national differences in policies that concern women's participation in non-traditional occupations, and gender mainstreaming policies on an international level. The explanatory model is developed by relying on previously proposed models and conceptual categories developed by military sociologists that have most actively engaged in this research. This book offers the exploration of an understudied yet academically and policy-relevant issue and a new theoretical model that explains patterns of gender integration meant to address the imbalance between women and men in the military. It emphasizes the need for a more theoretically oriented understanding of gender as it entails the identification of the need for such a process, identification of domestic actors inside and outside government structures that are involved in subsequent policy design, implementation, monitoring and evaluation, as well as international security context, international organizations and norms that influence that process.

No other scholarly work has done that in political science. It problematizes the realist state-centric analysis devoid of individuals and domestic political actors and processes, as well as the lack of studies of the impact that international organizations' gender mainstreaming initiatives, military technological advancements, the changing nature of warfare and new non-contiguous battlefields have on states' gender integration policies, and in turn, on the increasing number of women to serve in the military. Secondly, it sets out to clarify and operationalize concepts, and situate both the discussion and terminology within theoretical literature in comparative gender policymaking and international relations, particularly feminist security studies. Thirdly, although prior studies in military sociology provide us with umbrella categories, this book will seek to add necessary deductive rigor and test variables both quantitatively and qualitatively to eliminate a number of proposed hypotheses. This would then leave us with a more concise and parsimonious model. Fourthly, unlike all previous analysis, this book expands the study to include NATO's new member states of Eastern Europe in the sample. More than a decade after becoming full members of the alliance, the lack of analysis of gender integration in Eastern European states has left us with a partial understanding of the policymaking process and the factors behind it. And lastly, by being the first project to empirically test international level variables, it offers a roadmap for the future analysis of gender mainstreaming in the military.

Defining Concepts

While a number of scholars have developed remarkable and seminal ways of analyzing gender issues in the military, there seems to be little conversation among scholars. Not only is the policy change neglected by the scholarship, but there seems to be little shared theoretical understanding of the subject and even less consensus regarding concepts, methodologies or even significance of questions posed. This lack of consensus has resulted in conceptual stretching, ambiguity, and methodological confusion. We end up having separate conversations, presenting papers and articles in separate and different roundtables and panels at academic conferences without much overlap and much needed scholarly debate and exchange that would allow us to place gender, women and military within the same theoretical and analytical framework. Yet, stable and a shared understanding of concepts and methodology provide a foundation of any research community. This book is a contribution to the field of security studies that have long neglected women and gendered perspectives. Stephen Walt defines *security studies* as "the study of the threat, use, and control of military force ... [that is] the conditions that make the use of force more likely, the ways that the use of force affects individuals, states and societies, and the specific policies that states adopt in order to prepare for, prevent or engage in war" (1991: 212). This most commonly cited definition of our field clearly demonstrates that security is not only about the use of force, but also about the way national governments plan to recruit military personnel, and ensure readiness and sustainability of that force. Studying gender and women in the military is therefore an essential, but overlooked part of that planning and policymaking project.

This is a comparative *military personnel policy study*, which means that it will aim to study states' "course of action or inaction pursued under the authority of government" (Heidenheimer, Heclo and Adams 1983: 4) regarding gender integration. Unlike in any other profession, military recruitment and personnel policies are entirely monopolized by the state (Huntington 1957) and no other government institution controls lives, practices, images, and policy process as the military does. While bullets and wars do not discriminate, it is the state that clearly does, however, by holding a monopoly on the decision over who gets to hold the gun and protect the state as a legitimate member of the military forces. Historically and traditionally, the military has been closely linked to gender stereotypes, and states have often deliberately fostered a relationship between masculinity and military. Hence, it is impossible to study gender integration without studying the military as a hierarchical and traditionally masculine organization that stands in direct opposition to gender equality principles espoused by modern democratic societies. For much of our history and in most cultures, the responsibilities of a soldier have taken on *"gendered"* attributes in line with societal expectations regarding appropriate roles for women and men. This gendered definition of a soldier is constructed and learned, both within our societies and in the military, and is not defined by one's biological sex.

The semantic problem is that males and masculinity, and women and femininity are confused and conflated, and the assumptions and expectations regarding the association between them are often wrong. Both women and men can have masculine and feminine characteristics and traits (Disler 2010), yet military ties to masculine characteristics are often perceived as antithetical and even hostile to feminine characteristics regardless of temporal or regional focus.

This book builds on the previous theoretical models studying women's military participation (WMP), as it adopts some of their measurements and variables, for the sake of parsimony and consistency. Carreiras argues that we ought to "distinguish between those factors that concern women's simple presence in the system, which are not responsible for relevant variation in the qualitative status of military women, and those referring to women's qualified presence, which instead seem to influence gender integration to a reasonable extent" (2006: 127). Therefore, this book does not focus only on simple ratios, women's percentages, and numerical representation, but rather on what she terms *gender inclusiveness*. I define this as the degree to which individual states have taken action to erase gender stereotypes, including passage of legislation, policies and programs to provide women with the same opportunities, rights and obligations in the military as men.

The measurement of the main dependent variable is amended to reflect policy changes since the original concept operationalization. The analysis will focus only on the political process of gender integration, and not on personnel interaction, soldier attitudes and "interpersonal dynamics" prior to or after gender integration policy has been legally passed by the state's authorities.[5] In other words, the study will not evaluate the effect that these policy changes might have had on the overall armed services, their effectiveness, unit cohesion and readiness. Although "policy" is to be distinguished from "problems", meaning that state-sanctioned rules and regulations that govern employment, and address the aspirations and needs of women and men in the military forces, are not the same as implementation issues, this book also seeks to show that the two are often influenced by the same static views on the masculinity of the military profession.

Another term that is being increasingly used by policy scholars, policymakers and advocacy groups around the world is *gender mainstreaming*. Although the question of "mainstreaming" or the strategy of situating women's issues in the middle of a larger international debate on development has been around since the mid-1980s (Razavi and Miller 1995; Parpart), the term itself became part of policymaking and policy scholarship after it was presented in the Platform for Action of the Fourth World Conference on Women in Beijing (Pollack and Hafner-Burton 2000). This study will therefore adopt the official terminology offered by the Assistant Secretary-General and Special Adviser to the Secretary-General on Gender Issues and the Advancement of Women, whose responsibility is to support and monitor the implementation of it throughout the United Nations system. According to them, *gender mainstreaming* is:

---

5   For the microlevel analysis that seeks to explain how soldiers feel about the gender integration, discrimination and civilianization of the military see chapter 7 in Carreiras (2006) or Titunic (2000: 229–257), Miller (1997: 32–35), Stevens and Gardner (1987: 181–188).

a globally accepted strategy for promoting gender equality. Mainstreaming is not an end in itself but a strategy, an approach, a means to achieve the goal of gender equality. Mainstreaming involves ensuring that gender perspectives and attention to the goal of gender equality are central to all activities—policy development, research, advocacy/ dialogue, legislation, resource allocation, and planning, implementation and monitoring of programmes and projects.[6]

The need to familiarize ourselves with this term also arises from the recent integration of gender perspective by Security Council Resolution 1325 on women, peace and security that clearly requires more women to enter both peacekeeping and peace-support operations. The Committee on Women in the NATO Forces (CWINF) that since 1976 has been in charge of offering and sharing information regarding the status, organization, conditions of employment and career possibilities for women in the armed forces of NATO member states, has even changed its own name during the writing of this book, to the NATO Committee on Gender Perspectives to reflect its implementation of Resolutions 1325 and 1820. By changing the name, the alliance has shown its own commitment to the development of more effective ways to utilize women as a way to "promote, facilitate, support and monitor incorporation of gender mainstreaming in all NATO operation activities" (2003: 34).

While in recent times we have witnessed an increase in irregular forces, paramilitaries, rebel forces (Kaldor 1999), and corporate warriors (Avant 2005; Singer 2003), it is necessary to understand that this study will only analyze states' military forces. According to Max Weber, *the state* is a human community that has a monopoly on legitimate violence over a specific territory (Caforio 2003). This book will adopt Weber's definition of the state and its use of the military as the most tangible expression of that monopoly:

> Next, it is necessary to clarify the term *professional military*. It will refer to those who pursue a lifetime occupational career of service in the armed forces, where to clarify as a professional, he [and she] must acquire the expertise necessary to help manage the permanent military establishment during period of peace and to take part in the direction of military occupation if war should break out." (Rukavishnikov and Pugh in Caforio 2006)[7]

Therefore, this term assumes a long-term commitment of professional individuals who are not conscripted and who are not participating in military operations on

---

6   This definition is available on the website of the Office of the Special Adviser on Gender Issues and Advancement of Women Department of Economic and Social Affairs. Available at: http://www.un.org/womenwatch/osagi/gendermainstreaming.htm (accessed November 16, 2008).

7   Vladimir Rukavishnikov and Micheal Pugh have adopted this definition from Sills (1972: 305).

behalf of private military corporations. Before proceeding further, however, it is also important to mention that other terms such as armed forces, armed services or military forces will be used interchangeably to describe a professional military.

Argument

Analysis and argument are divided into four explanatory categories on two different levels of analysis: system and state. The three categories on the state level are military manpower, domestic political and economic factors, and cultural factors. On the system level, analysis examines the impact that international factors have on the state's behavior toward the integration of women soldiers. While it relies on the previous models for these categories, and tests them empirically, it argues that not all are useful in predicting greater gender integration. The main finding of this book is that civilian policymakers and military leadership no longer surrender to parochial gendered division of the roles, but rather integrate women to meet the recruitment numbers due to military modernization and professionalization; to meet the demands of domestic women's movements and to meet state's responsibilities under international agreements regarding gender equality and gender mainstreaming in the military.

More specifically, the first cause of the policy change is that of military personnel supply and demand in the twenty-first century. Integration and expansion policies are increasingly implemented in order to fill ranks due to the abolition of draft requirements. Similarly, when military battlefields move from the ground to computerized rooms thousands of miles from the front line, when better weapons and communications modes are available, and when an individual's physical strength is no longer the prime criterion for military prowess, the greater the gender integration. Technological improvements also create the need for specialized knowledge; therefore by extending the recruit pool to include both sexes, military forces have a greater chance of filling their ranks with experts. Also, the book shows that greater gender integration will be achieved when the levels of women in professional and technical fields are high in both private and public sectors. Great myths of masculine soldiering have been buckling under the weight of a new type of security environment that requires technical mastery and professional skills.

Secondly, in terms of domestic political context, the study demonstrates that numbers of women in the legislative branch and on a ministerial level are irrelevant to the passage of laws on gender integration in the military, but that the presence of liberal and autonomous second wave women's movements outside of government working together with women in government might help create a "strategic partnership" and open the window of opportunity for women in the military. This is particularly true in Western European and North American societies with a lengthy tradition of gender equality in other occupations and policy areas that have allowed women's movements to access a normative, legislative analytical framework for promotion of greater gender integration in the military. This is less true in Eastern European states where such autonomous women's movements within the context of the legacy of former regimes and transition process were largely nonexistent and irrelevant in legislative debates.

Thirdly, unlike authors who have previously attempted to answer the same question, this book finds that cultural factors are not among the primary determinants of defense personnel policy in the twenty-first century. While it is not excluding cultural factors from the theoretical model, the book shows that issues of national security do not allow policymakers and military leadership to surrender to traditional gendered division of the roles. Societal definitions and perceptions are less important than overriding military organizational needs and the state's international security commitments.

Furthermore, this book argues that current debates tend to ignore international organizations as sources of gender mainstreaming norms in military forces. Increased pressure by the United Nations to integrate gender perspectives into security and NATO seeking standardization and consistency are contributing to greater gender integration in the military. Again, there is a significant difference in the impact that NATO has played in new and old member states. While evidence shows that the alliance encourages gender equality in its ranks, as states admit women around the same time those same states are admitted into the alliance, admittance alone does not immediately translate into a high degree of inclusiveness. This is particularly the case among the new states of Eastern Europe, where the primary reason for the opening of ranks to women lies in the processes of democratization and accession to NATO, as they both demand more balanced gender relations. However, the degree of that integration in Eastern Europe depends on other factors, such as the transition to an all-volunteer force, and fewer offensive military capabilities.

Method and Methodology of the Book

The study combines both quantitative and qualitative methods in order to offset weaknesses of each and strengthen the value of the study. The mixed method allows for the triangulation of the data and will allow me to verify and contextualize the findings (Tashakkori and Teddlie 1998; Tashakkori and Teddlie 2003; Cresswell and Clark 2007). It will improve the overall quality of the research done, and show that the two approaches are not mutually exclusive but rather complement each other.

Quantitative Analysis

Quantitative analysis allows us to test a large number of variables in a large number of states. This method has not been used by scholars of comparative military personnel policies. Most published works explore individual states without much comparison or factor analysis (Murray and Viotti 1994). The case is rather similar with the students of comparative gender policymaking (Weldon 2002). In fact, there has been much disagreement over whether it is necessarily good to conduct large-N quantitative studies of gender policy scholarship because, as some suggest, statistics are associated with "masculinity" and therefore the scientific method suffers from gender bias (Keller 1983). There have been more recent feminists studying international relations who argue that by ignoring quantitative methods "feminists marginalize and devalue the applicability of quantitative research for furthering feminist goals and, ultimately, themselves as well."[8] In fact, in recent years there have been several large-N quantitative analyses that have helped advance our understanding of the processes that bring about gender equality in society (Gornick, Meyer and Ross 1998; Weldon 2002; Inglehart and Norris 2003; Caprioli and Boyer 2001).

By using the quantitative method, this study will seek to establish which one of the factors tested has more impact on gender inclusiveness in military forces cross-nationally. It will conduct the analysis by running multiple linear regressions and correlations using SPSS software. I will use the SPSS Pearson correlation to measure the strength, direction and significance of the relationship between two variables. Correlations tell us whether two variables are linearly related to each other. Multiple linear regressions are used to examine the relationship between my dependent variable, gender inclusiveness index, and all other dependent variables. Regression not only goes beyond correlation by adding prediction capabilities but it permits statistical control on confounding variables. In addition, scatterplots will be used to show the relationship between two or more variables. They help illustrate the direction, strength and form of each relationship.

Process-Tracing Case Studies

Quantitative analysis is extremely useful in identifying the most important factors that influence the process of integration, and it allows us to eliminate variables from the previous "laundry list" models. However, in order to see if the general arguments reflect the process of policy change in individual states and to better observe and contextualize human behavior, it is necessary to explore specific cases of gender integration. Here I used the process-tracing case study procedure to identify a causal sequence of events to illustrate how each my independent set into motion a gender integration policy change in different states. This process is different from a simple historical description as it provides a genetic and sequential explanation of gender integration policies, embedded in the theoretical literature

---

    8   Caprioli (2004: 253). Caprioli's argument is directed at those who openly discard or simply fail to recognize feminist empirical works in international relations because in their opinion is based on sexist assumptions and methodology. She accuses them of hypocrisy because by rejecting quantitative works, they themselves create hierarchies—something that feminist IR scholarship as a whole has been trying to deconstruct.

and empirical correlations presented in previous chapters. The method provides richness in detail as it allows for a closer scrutiny of individual and organizational actors on the domestic level of analysis.

Studying how different sets of variables affect the policy process in the most responsive and the least responsive states will help enrich the overall research. It also helps address a consistent criticism that quantitative analysis can give us correlations that mean very little in the "real world." Consequently, this book will seek to paint a better, more detailed and more concise picture of gender integration in the military.

The case study chapters will focus on four very different NATO members: the United States of America, Italy, Hungary and Poland. These were chosen based on two different but equally important variables. The first is the degree of gender integration in the military. Both Hungary and the United States rank highest among all the states, while Italy and Poland are the lowest. The second variable is the timing of NATO accession as it explains the difference in the timing of gender integration policies between Eastern Europe and Western European and North American states, but not the degree to which states have completed their promise to the alliance. Hungary and Poland are both new members, joining in 1999, whereas Italy and the United States were its founding members in 1949.

Focus on NATO Membership and Democracy

This study focuses on 24 of the North Atlantic Treaty Organization's (NATO) 28 member states. Iceland has no standing army and Estonia had not submitted a single yearly report on gender integration with NATO's Committee on Gender Perspectives by the time this book was completed, and therefore both have been omitted from the analysis. In addition, once the data-collection process was completed and the analysis of data finalized, Croatia and Albania were welcomed into NATO. Hence, these four states have been left out of the study.

Why NATO states? Firstly, the sample units are culturally, socioeconomically and militarily diverse and the evidence from this large-N empirical study will help make generalizations on the relative significance of the factors. About half of the states are Catholic, whereas the other half are Protestant, and Turkey is the only Muslim state. GPD per capita varies from $85,000 in Luxembourg to $12,500 in Romania. Their military expenditures also vary greatly, with Turkey being the biggest spender (5.4 percent of their total GDP) and Luxembourg the lowest (0.9 percent of their total GPD).[9]

Secondly, NATO member states offer information-rich cases because no other military alliance systematically compiles data or requires its members to submit

---

9 CIA Factbook Country Comparison GDP per capita (PPP) and Military Expenditures tables are both available at https://www.cia.gov/library/publications/the-world-factbook/docs/rankorderguide.html (accessed March 5, 2009).

yearly reports regarding gender integration within their armed forces. This allows me to obtain reliable data, access and conduct interviews with military officials and then test a large number of hypotheses on both a domestic and international level of analysis.

Thirdly, from a theoretical perspective the sample allows for closer scrutiny of two sets of categories. One is the proposed set of hypotheses regarding women's participation in politics, civil society and labor markets. According to Weldon, it is important to study gender policymaking in democratic states, because it is possible to measure "the degree of latitude afforded women's organizations" (2002: 22) and therefore their ability to organize, and work within and outside, government structures. States that are not considered democratic do not necessarily allow for the same level of women's participation in the government, or the necessary freedoms of speech, assembly and association. They tend to impose strict regulations on the formation of political parties, their activities and their agendas. In such states, public opinion can be manipulated by government authorities, leaving little or no space for gender equality advocacy. Thus, it would be difficult to assess what role non-governmental organizations, women's movements, legislators and political parties play in integrating and expanding the role women play in the armed services.

All NATO member states are considered electoral democracies by the 2008 edition of Freedom House's annual report.[10] Of 24 states considered here, 23 are classified as "free" states. This classification means that besides having democratic electoral systems these states respect and protect civil liberties, have no corruption, support a free and independent press, and nurture the free associational life of their citizens. Only the Republic of Turkey is classified as a "partially free" state. It held free and fair parliamentary elections, but civil and political liberties remain somewhat limited, corruption is widespread, and journalists practice self-censorship in order to avoid discussing sensitive topics.[11] Although Turkey is not as free as other member states, I have retained it in my sample in order to see if it

---

10   In order to be classified an electoral democracy, a state must have met the following criteria: (1) A competitive, multiparty political system; (2) Universal adult suffrage for all citizens (with exceptions for restrictions that states may legitimately place on citizens as sanctions for criminal offenses); (3) Regularly contested elections conducted in conditions of ballot secrecy, reasonable ballot security, and in the absence of massive voter fraud, and that yield results that are representative of the public will; (4) Significant public access of major political parties to the electorate through the media and through generally open political campaigning.

11   Journalists can be sued, arrested and jailed under the penal code for writing on Armenian genocide, division of Cyprus, and insulting Ataturk's principles of "Turkishness" and the sanctity of military forces. More information is provided by the Reports sans Frontiers organization which publishes an annual report on freedom of the press around the world. Available at: http://www.rsf.org/article.php3?id_article=25503 (accessed March 5, 2009).

will offer any additional answers or even generate additional questions regarding the integration of women into the military.

Lastly, one of the main arguments in this book is that norms regarding women in the military are disseminated through international organizations such as the United Nations, the European Union and NATO. Studying these 24 states allows me to evaluate the gender mainstreaming and standardization initiatives within these international organizations and what, if any, effect they have on the change of domestic policy on women in the armed forces. We can then seek to identify the mechanisms through which these norms trickle down to individual states. This research would then be useful to students and scholars of comparative politics, the military, and gender studies. Moreover, it would allow us to study theories of international relations regarding actors involved in the designing and implementation of national security policies, and in particular the role that international organizations play in disseminating information and encouraging policy innovation and formulation at the domestic level.

Outline of the Book

In this introductory chapter, I have sought to firstly present the puzzle, the dramatic change in policies regarding the gender integration and subsequent expansion of women's roles in the armed services in some states and stalled progress in others. Chapter 2 reviews the academic scholarship that raises fundamental questions about security theories, concepts and assumptions used to analyze military forces in international relations in general, and specifically the integration of gender perspectives in military forces. This chapter will demonstrate that it is precisely what Tickner eloquently termed "awkward silences and miscommunications" among different academic approaches that has left the examination of integration policy out of both gender and security academic writings and journals (1977). The theoretical discussion firstly addresses traditional security scholarship. It also presents the current conversations among critical security approaches, and in particular the broadening of the security agenda to include "securitization" and human security approaches, which continue to omit gender from their analysis. Secondly, it will concentrate on feminist explanations of women's military participations that help in framing a theoretical approach to understanding gender integration. Next, the section presents the military sociology as the only academic field that has attempted to build theoretical models explaining women's military participation. It will review and assess their conceptual, empirical and methodological approaches, theoretical contributions as well as inconsistencies to demonstrate the need for the theoretical model reevaluation and reconstruction. The second part of this chapter offers appropriate remedies and develops a theoretical framework for cross-national gender integration in the military and policymaking research. It clarifies the concepts and presents the arguments by drawing on security and gender literature that was neglected by the previous models, and seeks to situate the policy discussion within a larger theoretical framework within both comparative politics and international relations. In this chapter, analysis and argument are divided into four explanatory categories on two different levels of analysis: system and state. The three independent variable categories on the state level are: military manpower supply and demand, domestic political and economic factors, and cultural factors. On the system level, analysis will examine the impact that international variables have on the state's behavior toward gender integration. It will also show that we should not use aggregate data on women that bunch together women's political, economic and social activities into one measure. Instead, it will clarify how each individual activity has an impact on gender inclusiveness in the military.

By extrapolating data from state reports on gender integration and surveys conducted with the female officers and delegates to the NATO Committee on Gender Perspectives, Chapter 3 first develops and defines the six indicators based on which the dependent variable—gender inclusiveness—will be measured in this book. It presents the four categories of hypotheses, and empirical findings of the large-N quantitative analysis. The multiple regressions demonstrate that gender integration in the military can be explained by the abolishment of conscription, professionalization of the military services and technological advancements that require soldiers to have technical skills and not necessarily be present on the battlefield. In addition, the research shows that gender integration is more extensive in those societies where there is greater gender integration in the economic sector measured by female economic activity and women's presence in technical and professional fields, and where there is the presence of strong women's autonomous movements. The unexpected finding is that there are different factors explaining gender integration in new (Eastern European) and old NATO (Western European and North American) members. The timing of a state's NATO accession explains the timing of the admittance of women, but the length of membership is not correlated to levels of gender integration. Moreover, in "new" member states gender integration levels are greater in states with lower operational capabilities. The quantitative analysis also demonstrates the need to study the subject qualitatively, as it fails to answer some of the research questions, particularly regarding these regional differences.

Chapters 4 and 5 are case studies of the United States and Italy, and Hungary and Poland, respectively. These chapters trace the policy process in four different states and confirm that, in fact, there are different forces at work in the original member states of NATO in Western Europe and North America, and in the new member states in Eastern Europe. Chapter IV presents the evolution of the roles played by women in the United States, from the early days of the Women's Army Auxiliary Corps during World War II, expansion of the participation in the 1970s, and culmination with the wars in Iraq and Afghanistan, as part of the so-called War on Terror. Similarly, it will trace women's military participation in Italy from the early days of partisan guerrilla warfare and fascist women's corps, their return to the private sphere and home, and Year 2000 integration in the modern Italian armed forces. This chapter will show that besides the abolition of conscription as a determining factor behind the policy change, we need to look at the role that liberal women's movements have played in advocating greater gender integration in the original democratic member states of NATO. Secondly, it is demonstrated that the level of gender inclusiveness in the military in the United States and Italy are largely affected by women's quantitative and qualitative presence in labor markets, clearly pointing to the change in the operational environment that requires highly trained and educated soldiers, regardless of gender. Chapter 5 traces the process of policy change in the newly democratized states in Eastern Europe who joined NATO only in 1999 and integrated women as part of a major overhaul of their military services. Despite women's active participation in anti-fascist and resistance movements in both states during the world wars, female soldiers were few and far between due to the conventional mass military model that required the participation of all able-bodied men. The fall of the Soviet Union, the dissolution of the Warsaw Pact and democratization open doors for a serious discussion of gender equality, discrimination and equal opportunity in the labor sector. The integration in both states intensifies with the abolition of conscription and the move to an all-volunteer force. The analysis demonstrates that women's socioeconomic and political status have not played a determining role despite being on a par with their Western European counterparts. Similarly, in Eastern Europe, domestic women's movements have not actively engaged in a policy change nor have they been involved in the decision-making process to the extent that their Western counterparts have. Rather, it was the very democratization that banned labor discrimination, combined with the gender mainstreaming policies of NATO itself, that seemed to have been a greater incentive to incorporate gender in Eastern European militaries than in the West. In addition, the chapter will explore the unique security characteristics of these newly ascended NATO members, such as the level of military operational capabilities that might help us understand the difference in the level of gender inclusiveness in Hungary and Poland.

Chapter 6 summarizes all the findings, and makes policy recommendations particularly in regards to the expansion of women's role within NATO in the future. Although this book is not meant to promote one particular side or argument, it sets out to demonstrate the need for further gender integration, based on manpower and technical personnel shortages that most of the alliance members will encounter in the years to come. Next, the conclusion will briefly suggest future avenues of research in the field, particularly regarding the expansion of the study to new members and future candidates in the Balkans. As these states prepare to join NATO, they are in the process of revising their personnel and gender equality policies to conform to both the alliance and UN Security Council Resolutions 1325 and 1820. The conclusion then argues that it is necessary to study democratic states elsewhere in the world that integrated their ranks without having been members of the military alliance. It recommends comparison with non-democratic states, and it poses questions regarding the future of the military profession in general, as the state's monopoly on national defense and military services is being broken up by private military corporations and the effect that the outsourcing of military functions, operations and services will have on the role that gender plays once the society becomes only a client, and no longer a provider of national security.

Chapter 2

Gender Integration in the Military: Conceptual and Theoretical Framework

In this chapter, I seek to place the discussion on women and gender in the military within historical and theoretical contexts. A brief historical overview of women's military participation reveals masculinity as a recurring theme in the construction of "soldiering" and the importance of gendered ideologies in the military. Military force is presented as a male dominion and therefore closely associated with masculinity. These historical accounts illustrate that masculinity is considered the norm, produced in opposition to the other, femininity, which becomes an undesired characteristic and an exception in the military. As such, women's participation is overshadowed and marginalized by the heroic achievements of their male counterparts. The historical gendered perceptions of the military highlight how such representation of gender in the armed forces erases women's military identity and limits them to subordinate and supporting roles. This overview exposes much about the role that masculinity plays in the construction of soldiering, and by extension, about theoretical context that relies on the military institution as a central component of its conceptualization of power and war. In fact, military masculinity continues to inform the way we conceive of and study security. The argument here is not to suggest theoretical conspiracy against women soldiers, but to draw attention to the holes in our knowledge on what determines the degree of gender integration in the military. Traditional security studies claim that states recruit women as a "human resource" to satisfy manpower needs for the protection of the state, while feminists argue that such analysis obscures the gendered nature of that recruitment and explain state policies in terms of the patriarchal exploitation of society's resources. Both ignore that there are differences in the way states respond by demands of domestic political actors and movements seeking gender equality in democratic states, and international security organizations requiring the incorporation of gender perspectives into the military. While understanding of military masculinity is useful, as it foregrounds historical and theoretical discussions, I suggest it also ought to be understood as a dynamic concept tied to technological, economic and social transformations that no longer allow for a static view of a manly warrior.

Military Masculinity and Women Soldiers: Historical Debates and Controversies

Although our military strategy studies date back to early analysis of the Peloponnesian Wars and Sun Tzu's writings on war, one would have a very hard time finding any references to women who have actively participated in the events they describe. According to Plutarch, Spartan warrior princess Arachidamia "with a sword in her hand" entered the Senate and demanded women be allowed to fight Pyrrhus's invading troops. She and her women were pivotal in Sparta's successful defense of the besieged Lacedaemon. In ancient China, to demonstrate his war-making abilities and put his own skills to the test, Sun Tzu trained 300 concubines of King Wu and turned them into loyal soldiers. Hence despite their presence on battlefields for more than 3,000 years, women soldiers remain an understudied, unexplored and undervalued subject in social sciences. Despite the plethora of archive resources and documentation that has allowed modern historians and social scientists to gradually reconstruct the paths and activities of women in the military, they are still seen as an anomaly, lacking the respect and honor they well deserve regardless of cultural and historical settings. Instead, women continue to be portrayed as victims of men's militaries and never its devoted participants. It was pointed out more than a decade ago that gendered war roles are almost completely neglected by North American, mostly male security writers (Goldstein 2001). Instead, for the most part, in Western traditions, as Bethke Elshtain argues, men and women have been simplistically represented in terms of one-dimensional gender roles, and "in time of war, real men and women—locked in a dense symbiosis, perceived as beings who have complementary needs and exemplify gender-specific virtues—take on, in cultural memory and narrative, the personas of Just Warriors and Beautiful Souls" (1987: 4). And what this does is create a myth of a woman as a peace-loving and nurturing being, and men as chivalrous warriors protecting their non-combatant women. Men are supposed to join the military and take up arms to demonstrate their "masculinity" and prove their manhood. They are admired for that, while women doing the same thing, defending their countries, have received a different, much cooler reception by society. Similarly, Tickner points out that such war stories "rely on the portrayal of a certain kind of masculinity associated with heroism and strength ... Rarely do war stories include stories about women" (2001: 57). Joining resistance efforts was often portrayed as "unwomanly" and motivations are described as emotional and unreasonable responses to loss or scorn. This version of gender discourse really meant that women's military and political identity was going to be defined in terms of differences between women and men, rather than being based on women's military contributions. As Dowler points out, historically, there is a "tendency to perceive men as soldiers, warriors and heroes of war, while women are understood as the victims or icons of ... war" (2002: 161). Such a discourse has created a gendered perspective on the military; on the one hand, the man's loyalty to the state, his citizenship and even economic future were often inherently tied to his military service, while women are only to be used by militaries to dehumanize, humiliate and victimize the nation. Hence, women were limited not by their biology but by societal expectations. It is precisely because of such gender expectations that women's contribution and activities have been deliberately overlooked by historians. Even women soldiers have downplayed their own roles, despite the overwhelming evidence, as they felt that men's military efforts were real, and women were only there to support their men (Alfonso 2009). This only further preserved and propagated the societal notion of manhood and masculinity as chief components of the official construction of military identity, and where women were only in the subordinate, supporting and traditional roles associated with femininity.

It is true that for centuries it is men who made up the majority of recruits and held the highest ranks, although women were there, too. Gendered roles in the military are only patterns of behavior that humans are expected to conform to, and as Elshtain points out, are not representation of realities, particularly not of war. While young girls may grow up reading of princesses needing to be rescued, very few know the stories of Queen Dahlia of the Moors commanding Barbarian forces against Arabs, Nigerian Queen Amina, "a woman capable as a man" who led her troops in conquests for almost 34 years, or Vietnamese sisters TrungTrac and TrungNhi leading an 80,000-strong army against their Chinese rulers in A.D. 40. In A.D. 60, Celtic Queen Boudicca, who had been tortured and her daughters raped by the Romans, proclaimed:

> We British are used to women commanders in war. I am descended from mighty men! But I am not fighting for my kingdom and wealth now. I am fighting as an ordinary person for my lost freedom, my bruised body, and my outraged daughters ... Consider how many of you are fighting—and why! Then you will win this battle, or perish. That is what I, a woman, plan to do!—let the men live in slavery if they will. Tacitus, *Annals* (XIV. 35)

Although she led British rebels to many victories that made Emperor Nero reconsider his tactics, these were described as driven by anger and revenge, her strategy in Roman Britain cruel and brutal, and her troops as a nationalist, looting mob. Joan of Arc was a child of peasants who became a ferocious leader, placing herself at the head of every assault, regardless of danger and her own wounds. Despite enormous historical evidence, the fact that such an unlikely character changed the course of war, her military leadership role is still deprecated and Joan portrayed as no more than a cheerleading inspiration mascot with a flag. During the American Revolutionary War, a few courageous women including Margaret Corbin and Mary Ludwig Hays McCauley, known as Molly Pitcher, served in combat alongside their husbands and others such as Deborah Sampson disguised as a man to serve in the Continental Army, against every social convention of the time. Their military combat experience and service in that war are well documented as they were among the first female soldiers to earn military pensions. Yet, their achievements are often secondary to the descriptions of other types of contributions in which women did not step out of traditional feminine roles associated with the domestic sphere, such as the women-led boycott of British goods in the 1760s and early 1770s, and women's production of food and clothing to support the war efforts. In the past decade, historians DeAnne Blanton and Lauren Cook Wike at the National Archives have combed through diaries, letters, burial records, military reports and newspapers documenting the service of women soldiers, and found that an estimated 400 to 1,000 women, or more, disguised themselves as men and fought during the Civil War (2002). This evidence not only proves that women in the military is not a new phenomenon, but also one that was almost lost to history as the reality of these women's lives failed to conform to gender stereotypes.

Although it is said that World War II had a huge impact on women's status in all societies, women's resistance activities around the world continue to be forgotten. Kirk and McElligot point out: "Armed women are either written out or deprived of their female identity, while women in nurturing or caring roles lose any claim to be equal resisters."[1] One of the most famous and feared regiments during World War II was an all-female 588th Night Bomber Regiment of the Soviet Air Forces (later renamed 46th "Taman" Guards Night Bomber Aviation Regiment). Dubbed by the Germans "*Nachthexen*," or Night Witches, they flew combat and offensive missions in Stalingrad, Sevastopol, Minsk, Warsaw and Berlin in outdated, wooden and open-cockpit Polikarpov Po-2s that could only fly at night to avoid *Fliegerabwehrkanone*, the German aircraft defense cannon. They flew more than 24,000 missions and dropped 3,000 tons of bombs as part of their harassment bombing and precision bombing missions from 1942 to the end of the war. Twenty-three women became Heroes of the Soviet Union, making them the most highly decorated female unit in the Soviet Air Force, yet the Soviet public, and even some at the highest levels of the military, were unaware of their existence. In the West, many dismissed the stories of the Night Witches and regarded them as propaganda tales of the Soviet regime. Although officially the Soviet regime treated women as equals, propaganda often promoted domestic sphere and family care as women's jobs, revealing contradistinctions of the so-called Communist emancipation of women. Despite the women's stellar military record, patriotic and military exploits were considered male realms. Femininity and the physical appearance of women soldiers were emphasized, as was their motherly and sisterly care for the fighting men (Cottam 1983; Conze and Fieseler 2000). The Soviet state ignored women's participation after the war as all-female units were sent home, and for the most part women were prohibited from entering military schools and ranks.

In recent years, we have seen a proliferation of military history works that offers considerable insights by documenting the historical presence of women in the ranks. DeGroot and Peniston-Bird offer a rich selection of essays exploring the role of women as camp followers and supporters of military forces during conflict

---

1   Kirk and McElligot, *Opposing Fascism*, 10.

and in peacetime before and after World War II in Europe, North Africa, Asia and even the Middle East (2000). Similarly, Linda DePauw reviews the role of women as warriors in all eras and in all parts of the world, from prehistory through the Crusades and revolutions to the Great War (1998). Others focus only on the involvement of U.S. military women in the Civil War (Harper 2004; McDevitt 2003) and in World War II (Gruhzit-Hoyt 1995; Saywell 1986). One only needs to look at the plethora of personal memoirs written by women soldiers who have recorded their experiences within the military structure, and biographical works about individual heroines from Panama to the Gulf Wars and Afghanistan that are a testament to the extent of recent policy changes (Churchill 1992; Cornum and Copeland 1992; Holm 1982; Walker 1994; Bragg 1996; Edmonds 1999; Adams-Ender and Walker 2001; Bragg 2003; Karpinski 2006; Wise and Baron 2006).

Despite this wealth of information documenting women's military participation, women are rarely heroes, and even less so perpetrators of violence. The notion of a hegemonic masculinity that has long permeated military culture excludes women from all except subordinate roles. Women's relationship with the military ends up being associated with two themes: feminine women as victims of military, and feminine women as scandalous and supporting protagonists in war, such as sexual workers, camp followers and treasonous spies. What we see is that no matter what part of the world stories are coming from, societies have "defined the soldier as an embodiment of traditional male sex role behaviors" (Barrett 1996: 129) and the military institution is represented as an embodiment of military masculinity. According to Woodward, "the warrior hero is physically fit and powerful. He is mentally strong and unemotional. He is capable of both solitary, individual pursuit of his goals and self-denying contribution towards the work of the team" (2000: 641). The physical fitness, discipline, bravery and aggression are all characteristics of hegemonic masculinity (Cheng 1999; Martino 1999; Connell 2005) as is emotional maturity and stability. Emotional vulnerability is considered a feminine attribute, and is seen as a threat to military morale and effectiveness (Mankayi 2006). Women are represented as frail beings, without the physical ability to fight wars, and unable to control their emotions. Soldier roles are for masculine men and if women are "soldiering," then they are treated as an unnatural disruption, an anomaly in the hegemonic and gendered military structure. As such the warrior is on the top of all societal gender hierarchies. Military masculinity subordinates all other cultural and societal gender constructs and requires femininities to exist, in order to produce this highly gendered organization. States have for centuries produced, reproduced and relied on military masculinity as an institutionalized form of exclusion of all other masculinities and femininities.

As has been the case throughout history, continued focus on the link between masculinity and soldiering has over time allowed for very little change in gendered structure and some argue, military has not really evolved as an institution (Dudink et al. 2004; Enloe 2000; Higate 2003). If femininity is antithetical to soldiering, masculinity remains as the only path to "real" soldiering. Recently, when a female student and Iraq war veteran described herself as a "mean lean killing machine," many students in my classroom found this construction of her military identity objectionable, because "killing" as a concept was not a feminine attribute. Such a practice preserves the hegemonic masculinity of the institution and discursive practices, as women are adopting masculine perspectives and norms, and abandoning femininity as they enter the military (Rimalt 2007). In fact, by attaining military masculinity, women are often finally able to legitimately claim their authority. But in doing so, some argue, women have helped obscure, legitimize and preserve this gendered system and military masculinity that continues to favor men, and is only further reinforced by the widespread and seemingly tolerated high levels of sexual harassment in the ranks (Belkin 2012). According to Herbert, in order to fit into a masculine military environment, women continually have to "camouflage" their femininity as that is "something to be denied or, at the very least, controlled" (2000: 45). Such a gendered ideology of the military institution makes the integration and policy process much more difficult.

The process has not been without its fair share of public debate and controversy, and has exposed the gap between the conservative and liberal writers, social commentators and policymakers. The gap seems to have widened following the terror attacks in the United States on September 11, 2001 and the wars of the Bush administration, as these dramatic changes in the international security environment, that now included non-state actors and irregular front lines, have contributed inadvertently and significantly to gender integration in the military. Yet the right to "soldiering" and state military personnel policies have remained largely a subject of ideologically charged debates among liberal and conservative pundits and war reporters, and been virtually unexplored by security studies. Women soldiers became prey to public imagination and sensational international headlines, polarizing, stereotyping and forcing us to choose between prisoner of war Pte. Jessica Lynch and rank-and-file torturer Lynndie England. Both served to perpetuate gendered myths about women in the military. Lynch was "a girl next door" in need of rescuing, England "a bad apple," and sexual deviant, portrayed by the media as a symbol of all that is wrong with gender integration while simultaneously serving as a scapegoat for crimes that were mostly committed by men in Abu Ghraib prison. The world followed Lynch's journey from the moment she was captured by Iraqi officers in Nasiriyah in 2003, saved by macho all-American warrior males, and returned to the safety of her home. And once again, "men were men because they saved women, and women were women because, once saved, they clung to their savior's hand and returned to the domestic fold."[2] The media hardly explored the ease with which she was saved, or her role as a soldier. Rather we read about Special Forces military men rescuing an all-American "young, cute, blond" female in need of protection from the foreign, brown evildoers. She was stripped of her soldier identity, to become a Disney-like fantasy princess worthy of our war, and requiring a remasculinization of the military. The public was presented with a gendered myth, of avenging, macho,

---

2   Faludi, 224.

hypermasculine "John Wayne style" soldiers, and a brittle, innocent girl in need of rescue. The whole episode served as a carefully scripted scenario in which masculinity and femininity inform one's place in the military and in conflict. There were two other women involved in the incident: Lynch's colleagues Lori Piestewa, a Native American single mother, and Shoshana Johnson, a black Hispanic single mother. Unlike Lynch, who did not fire a single shot, Piestewa and Johnson actually engaged the enemy in combat, and Piestewa lost her life while Johnson suffered gunshot wounds. As warriors, they posed a threat to picture-perfect privileged military masculinity and did not fit the post-9/11 world that also brought on the celebration of "manly" police officers, firefighters, and special operations soldiers fighting for their women at home. Women's efforts and the reality of gender integration in the military were once more ignored and reshaped to fit new political realities.

This conservative, "John Wayne style" masculinity was quickly at odds with liberal activist celebrations of such GI Janes, who fought and died in Iraq and Afghanistan in unprecedented numbers, and are still regularly sent out on missions with all-male combat units for the purpose of defusing tensions.[3] The U.S. Marine Corps has 16 Female Engagement Teams based in Helmand province, Afghanistan, while Britain has four.[4] The new wars exposed female soldiers to volatile conflicts without clear front lines, and by default, combat operations for which they have not been trained. Women known as "Team Lioness" ended up fighting alongside the Marines in some of the bloodiest battles of the Iraq war. Although liberal commentators cheered them on, and cited successes as a way of challenging and removing the current ban on combat, they were confronted by a loud conservative opposition claiming that insistence on gender integration in the military has disrupted the "natural" order. Critics such as Brian Mitchell have long argued that gender integration is akin to social engineering that only creates a less effective and less disciplined military force, despite the fact that The Women in Combat Task Force Study Group found that for each cohesion-building factor, women either had a positive effect on building combat effectiveness or had a neutral effect.[5] According to Mitchell, "it has been twenty years since women first forced themselves into the federal service academies, where they have shattered tradition, fractured morale and confused academies' purpose—which is to train combat officers."[6] Like Mitchell, others argue that scandals of the Citadel, the Virginia Military Institute, Tailhook and Aberdeen simply forced the Defense Department and military establishment to adapt and inadvertently overcompensate for their past mistakes, and ended up ignoring a negative effect woman will have on unit

---

3   Goldstein (2002).

4   NATO Committee on Gender Perspectives, "Engaging Women on the Frontline," 18 July 2011. Available at: http://www.nato.int/cps/en/SID-B0C60A3A-E92B72D4/natolive/news_76542.htm?selectedLocale=en (accessed June 2012).

5   Saimons (1992).

6   Mitchell 1998, xvi.

cohesion, combat readiness, morale and, ultimately, the rate of success in military operations (Gutmann 2000). While Mitchell's work, is not necessarily a result of scholarly research, as it often lacks scientific data, and opinion polls quoted are not referenced, it has, however, influenced more conservative policymakers who continue to block gender integration initiatives in the United States and elsewhere (Weinstein and White 1997). For example, former House of Representatives Speaker Newt Gingrich, who is one of the adamant and vociferous opponents of women in combat, often made comments that echoed Mitchell's arguments such as claiming that "females have biological problems staying in a ditch for 30 days."[7]

Some correctly call these debates dishonest, because both sides are exploiting gender and women in the military to score "culture war" points.[8] Yet, what these heated public debates demonstrate is that policies of gender inclusion in the military services are one of the most dramatic changes in the defense sector, and it demands much greater academic scrutiny so that we can move beyond these narrow, divisive and uninformed arguments that serve to sustain military masculinity and the exclusion of women from the military. This book dispels these oversimplified, superficial and sensational myths that frame gender integration as militarily unnecessary feminist concoction, inherently in conflict with societal and cultural mores. Instead, I show that such narrow military imagery has little to do with military manpower needs and the realities of modern warfare that demand greater integration.

Security Scholarship: Gendered Security, Gendered Military

While the military as a national security instrument has traditionally been studied by international relations, and in particular security scholars, its interaction with the society it claims to protect has been largely neglected. Traditional security and military studies have always been within the realm of political science, and as such have tended to focus on states' grand military strategy rather than individuals participating in the military. But an overview of traditional security theory's construction of military and power relations and critical and feminist security theorists' arguments are useful in understanding gendered international relations and assumptions regarding hegemonic masculinity as a required characteristic of a warrior. It helps to reveal how traditional security scholarship perpetuates the myth of a "man warrior" and inadvertently limits the theoretical space for a much needed discussion of gender integration.

During much of the Cold War, security studies narrowly focused on the means by which to pursue security and that largely meant external threats, and the use and control of military force. What became known as "traditional" studies of national security and the military became synonymous with pessimistic realist theoretical

---

7   Seelye (1995).

8   Solaro (2010).

assumption that international relations are inherently conflictual, and just as men seek to dominate each other, states fight wars against other states. Neorealist Kenneth Waltz added a new dimension by emphasizing the anarchic nature of the international arena within which unitary and rational states exist, formulate their security policies and struggle for survival (1979). Anarchy does not imply that states are operating in a chaotic and disorderly world, but rather in a system that lacks the centralized rule enforcement mechanisms.[9] Pursuit of national security goals as a way of ensuring state's survival, and conceptualization of military as a key tool of statecraft that alone is able to promote national security, became central to the international relations and security scholarship.

A system "characterized by security competition and war" (Mearsheimer 2001:30), realist scholars argue, is the reason why the primary and almost exclusive public policy's goal is to increase state's military security. As the field became more scientifically oriented by the 1960s in its quest to prove that politics is indeed a science, often mimicking economics and using their game theoretical and rational choice models, security studies largely continued to concern itself with states' strategic thinking and development of effective military capabilities that will allow them to maximize their relative power position over other states. Just about every security policy decision and outcome, particularly regarding military force structure was to be explained by the states' rational and unequivocal desire to survive in the anarchic self-help system. Simply put, the greater the military might and advantage one state has over other states, the greater its security. In addition, the security dilemma, or the situation in which an increase in one state's security results in a decrease of security for the other state, can cause the states to strategically adjust their military capabilities including personnel.

Realism is not concerned with domestic actors or public debates regarding the recruitment and personnel policies of that military, or the interactions and relations between the civilian society and the military institution. Hence, any changes to the military gender structure and personnel was considered devoid of the social context and domestic political debates within which political actors engage, reflect and make policy. Personnel policies should simply be a result of careful cost—benefit analysis and free of all cultural and societal mores. Hence, one can deduce that gender integration policy would be explained as simply states optimizing in an ever more conflictual world.

Despite the lack of overt focus on the individuals, one can argue that the systemic explanations favor masculine approaches to studies of war, doctrine, strategy and operations Although it claims gender-neutrality, for the most part, classical realist literature from the times of Sun Tzu, Thucydides and Machiavelli was seen as advocating gender domination for the sake of state preservation (Goldstein 2001). Those who do study individuals tend to write "great man" theories and "great

---

9   Although a realist concept, anarchy has found its way into game theory. For an extended discussion on different conceptualizations of "anarchy" within international relations literature see Helen Milner (1991).

man" histories such as Daniel Bynam's and Kenneth Pollack's article "*Let us Now Praise Great Men*" or Kissinger's "*Diplomacy*." With the exception of a couple of references to Margaret Thatcher by Kissinger, there are almost no women. Just as in all other historical representations of soldiers, these great men exhibit the same *hegemonic masculinity*, or the ideal masculinity in international relations characterized as rational, unemotional, independent men defeating other men, exhibiting feminine characteristics, such as concern for morality, compassion and nurturing. To gain and maintain power, states need not just any kind of a man but a particular type of masculinity reserved for the toughest and hardest military leaders, capable of dealing with security threats and war and is "not … normal in the statistical sense; only a minority of men might enact it. But it was certainly normative. It embodied the currently most honored way of being a man" (Connell and Messerschmidt 2005: 832). In security studies, this hegemonic masculinity also assumes martial nature of masculinity, that Clausewitz, father of security literature, described not only in terms of bravery, physical strength and manly vigor, but also indifference to suffering, self-command and spirit of boldness. Those who have "inflammable emotions," meaning those who exhibit feminine characteristics of lacking emotional strength are "of little value in war," according to Clausewitz (1873: 123–124). This conceptualization of a warrior genderizes both military and state behavior as the "strong" states become reflections of their masculine, security-seeking rational leaders, defeating other "weak" states, led by soft, and thereby feminine, men.

Traditional security scholarship's narrow focus on the military strategy of great men and the requirements of combat serves as a scholarly justification for the military ties to hegemonic masculinity and argument against gender integration in the military ranks. Moreover, there is very little recognition that dysfunctional civil—military relations can produce inefficient strategic analysis, incomplete policy recommendations, and by default, negative policy outcomes. Who serves in the military is a question of national security and of strategic importance, particularly in democratic states where the answer is formulated not only based on the state's strategic and manpower calculations, but also the demands of domestic actors, socioeconomic considerations, and, increasingly, international norms and ideas. Lack of recognition of gendered construction of the state and military, gendered presentation of power relations among the strong and weak states, and continued scholarly attribution of martial virtues to great men, adds to the continued conflation of national security and masculinity, and representation of women soldiers as being detrimental to both. As Sjoberg and Via argue, "privileging who and what is masculinized is inextricable from devaluing who and what is feminized" (2010: 18). Such "naturalization" of martial and hegemonic masculinity in the military and national security not only subordinates, but also dismisses and degrades women who serve their states. And finally, it demonstrates traditional security scholarships' failure to recognize that martial masculinity is not ahistorical, but rather fluid and evolving. This traditional construction of military personnel that favors self-discipline, physical toughness aggression, violence, and emotional control in Clausewitz's terms is at odds with today's military needs that include communication skills, technology and linguistic skills, organizational science and anthropology. Militaries around the world are seeking to adjust to the new realities by no longer being a place where "men are made," but a place tied to an economic market offering careers and knowledge of technology, with a more compassionate and caring image (Elder, Jr. et al. 2010). As we have witnessed in the recent wars in Iraq and Afghanistan, the ever-changing international security environment and military personnel technical needs can alter gender roles both inside and outside the military and reconfigure ideas about what is appropriate for men and women and prompt greater gender integration. This detachment from socioeconomic, historical, technological and international security arena changes, among other things, prompted a broadening of the research agenda in security, albeit without much too much emphasis on gender relations and women's exclusion.

One of the first to offer a strong critique of traditional security analysis in the early post-Cold War days was the *"securitization"* school hailing from Copenhagen. This approach is initially associated with Oli Wæver's characterization of security as a "speech act" (1995), where the framework is built around the concept of "securitization" or leadership's positioning through discourse on a particular issue, that once presented to the audience as an urgent and existential threat demands their immediate consideration, policy change, and actions that often override the existing rules and normal politics (Buzan, Wæver, and De Wild 1998). By assuming that it is the elites setting the tone and security agenda, and not problematizing women's invisibility and absence within such circles (Hansen 2000), "securitization" literature inadvertently accepts traditional security link masculinity and military as given, and by default renders questions of women and gender in the military marginal. The criticism is not only directed at the Copenhagen school, but also 'Paris" and "Aberystwyth," which became part of the "CASE Collective" (CASE Collective 2006: 444). Today there are some significant contributions on the United Nation's security language framing of women's rights and gender equality (Hudson 2009) and post-conflict de-securitizing language of the formerly securitized female combatants in Sierra Leone (MacKenzie 2009). But if "securitization" literature is meant to explain how security problems emerge, evolve, and dissolve (Balzacq 2010), then not enough research has explored the question of who is tasked with providing security, and in particular, what domestic actors are articulating the "securitizing moves" that are used to incite, defend and rationalize policies of exclusion and integration of gender, and how they might vary cross-nationally.

The *"Human Security"* approach, motivated and popularized by the 1994 Human Development Report of the United Nations Development Program (UNDP), moves the focus of analysis from traditional state security to an individual as the main referent of security policy. Although based on the notion of individual rights, some worry the approach regurgitates dominant debates by not deconstructing the term "human," which is historically exclusionary and gendered (Marhia 2013) and therefore, there is a "real danger of collapsing femininity and masculinity" (Hudson 2005: 157). Yet, it is within this "broadening" that some of the earliest feminist security scholars carved out their academic space within this approach, offering a critique of realist theoretical "top down" systemic analysis that ignores women's experiences, and rejecting empiricist methodologies and the notion that the social world and natural world can be analyzed in the same ahistorical, apolitical and objective manner (Tickner 1997).

Making Feminist Sense of Gender Integration in the Military

A recent analysis of intersections of gender and security scholarships demonstrates that "less than forty out of more than 5,000 articles in the top five security journals over the last twenty years explicitly address gender issues as a major substantive theme" (Sjoberg and Martin 2007: 2). While one can trace the evolution of *feminist security* thought from the 1960s feminist writers within peace studies seeking to redefine and broaden the concept of security and threat (Wibben 2011b), the actual, and very late arrival did not take place until the late 1980s with classics that laid out the foundational work by seeking to place gender in international relations and analyze the relationship between military and masculinity (Stiehm 1982; Enloe 1983; Reardon 1985; Elshtain 1987).What really creates the momentum for a feminist study of international relations were the three academic conference proceedings being published in the 1988 special issue of *Millennium: Journal of International Studies*, including Ann Tickner's critique of Hans Morgenthau's principles of political realism. This is when we really start to see new and more cohesive examples of how to develop and integrate a feminist epistemology in international relations and security theory. What followed very quickly was Tickner's seminal work, *Gender in International Relations: Feminist Perspectives on Achieving International Security* which initiates a new discussion by bringing women into what she describes as "an almost exclusively male domain," of international relations, as it included virtually no female writers and as it regarded the analysis of the relationship between gender and security entirely irrelevant to policy and scholarship (1992). While most would agree that women have been involved on battlefields around the world, their experiences have been largely neglected by international relations theories. All gendered variables are deemed unimportant by the state-centric security studies, and women's experiences and understanding of events in international arena were ignored. Tickner's critique of realism is centered on its narrow definitions of security that exclude economic and environmental threats, and assumptions regarding the behavior of states in the international system which are solely based on men's experiences, and hence, can provide us only with a partial understanding of international relations. As Annick T.R. Wibben notes, "within IR Tickner's work is the most common starting point for elaboration of feminist thought in security studies" as it sets the tone and presents questions and research agenda for future feminist inquiries in security studies (2011b: 4).

In the last couple of decades, feminist writers added to the security studies a new perspective, a gender-based perspective intent on exploring, and placing women within the realm of war, post-conflict peace-building and international security policymaking to demonstrate that gender as an analytical category is not as completely invisible in international security as traditional security studies would argue. Feminist international relations contributed to how traditional security doctrines and policies are gendered and how they are linked to gender inequality in military. As Sjoberg and Martin point out, "to characterize Feminist Security theory singularly is to imply that there is a feminist theory of international security" (2007: 5). In fact, that would be misleading, oversimplified, as well as an epistemologically wrong description of a rich body of literature. Feminist security writers come from a variety of different feminist theoretical schools and "there are some real differences in feminist scholarship—and these differences matter" (Wibben 2011a: 590), particularly when it comes to their analysis of gender in the military. There is no collective perspective on the subject and divisions are at times acute. Although most agree that masculinity matters in understanding gender integration in the military, they often disagree on what explains why states pass such policies.

Structural explanations are largely based on feminist security works such as Cynthia Enloe's 1983 *Does Khaki Become You?* which is among the first to analyze military masculinity and concludes with an assertion that much of the male-dominated international security system is dependent on the women's subjugation and marginalization, including in the military. In her follow-up work, *Bananas, Beaches and Bases*, Enloe adds that regardless whether we study banana plantations, beaches of the sex tourism industry, or the lives of base women and diplomatic wives, states' military agendas and male-dominated international politics could not survive without women's subjugation (1990). The concept of "militarization," that Enloe continued to explore in her *Maneuvers*, is a "package of ideas" that requires international security to be gendered, as a way of affirming that masculine men are the protectors of the protected, the feminine women. If the state's security is based on its military might, and militarized power relies on the participation of those it subordinates, then one can conclude that states design their national security policies and strategies in a way that allows them to "use women for military purposes only in those ways that will not unsettle the military's masculinized status" (Jensen 2005: 206). Hence, according to Enloe, gender integration is a military necessity of patriarchal structure. Governments militarize women as it is necessary process of the militarization of men and societies. "Military policy makers have needed women to play a host of militarized roles: to boost morale, to provide comfort during and after wars, to reproduce the next generation of soldiers, to serve as symbols of a homeland worth risking one's life for, to replace men when the pool for suitable male recruits is low" (Enloe 2000: 44). In the 2013 interview, Enloe explains that regardless of the obstacles removed and progress toward gender integration in the military, as long as we live in militarized societies, there is no genuine gender equality. While she recognizes the importance of opening up more roles in the military for women, Enloe argues that "nobody should imagine that making a military a more legitimate institution rolls back patriarchy."[10]

Although critical and postmodernist/poststructuralist feminists have not directly engaged in the identification of factors explaining gender integration, there is a similar understanding of the structure of the military order, and its patriarchal and masculine aspects. Critical feminist scholarship builds upon critical theory's commitment to understanding the world in order to try to change it and theoretical framework that connects material conditions, ideas and institutions in what Robert Cox terms the formation of "world orders." Feminists look at "ideational and material manifestations of gendered identities and gendered power in global politics" (Dunne, Kirki, and Smith 2010: 189) and demonstrate how our theories are linked with practice. Sandra Whitworth explores changes in understanding of gender in international organizations and how gender informs the various institutions (Whitworth 1994a; 1994b) and finds that the way women are addressed as "requiring special attention" within organizations reinforces particular views about women in the workforce, and legitimizes the treatment of women as not "real workers." In the military "special attention" translates into separate physical and weight requirements, and arguably legitimizes women's position in a supporting role, not doing "real soldiering." As such, women are perceived as not deserving same rights as men, in and outside of the ranks. Their experience with the military is the one of subordination and marginalization for the sake of national security goals (Moon 1997), not inclusion and equal participation.

Postmodern/poststructural feminists see "female" cast into the role of the Other, and hence reject metanarrative explanations of knowledge and experience that are male-constituted and male-dominated. Charlotte Hooper's poststructural analysis on the role that international relations play in the shaping, defining and legitimizing of masculinity and masculinities (2001), and Carol Cohn's analysis of defense-community language and sexual imagery used to construct a world of nuclear weapons (1987), are very useful in our understanding of the so-called "cultural" wars over integration. Moreover, such works emphasize that "gendered images saturate the discussion and promote masculine qualities as superior" (Sjoberg and Martin 2007: 26), and as such demand a closer examination of how policymakers, advocacy groups, or international organizations construct the discourse of greater equality in the military, as well as the discourse of indoctrination, training, military discipline and combat.

Liberal feminists argue that gender integration in the military is a sign of "gender equality and equal citizenship for men and women" (Rimalt 2007: 1098). This approach was triggered more than 200 years ago by Mary Wollstonecraft, who in her *Vindication of the Rights of Women* demanded full political, civil and social rights for women; largely responding to Jean-Jacque Rousseau's *Emile*,

---

10   Alex Stark, "Interview with Cynthia Enloe." March 13, 2013. Available at: http://www.e-ir.info/2013/03/13/interview-cynthia-enloe/ (accessed May 2013).

which argued that women should remain in the private sphere, and leave the public sphere to men. Liberal feminists continue to fight discrimination and exclusion from the ranks but it is, however, still debatable whether Wollstonecraft did agree with women bearing arms or if women's obligations as a mother prevented and exempted them from soldiering. She wrote: "for though I have contrasted the character of a modern soldier with that of a civilized woman, I am not going to advise them to turn their distaff into a musket though I sincerely wish to see the bayonet converted into a pruning hook" (Wollstonecraft 1792: 219). Yet, her claims against the restrictions of political power, and continued efforts of liberal feminists, contributed to gender integration in a number of states. For liberal feminists, equal participation in the military is a question of ensuring the equal rights and responsibilities of all citizens when it comes to defending their country (Stiehm 1982; Fenner and de Young 2001), and as such, in democratic states it should be explained by existing institutional laws, rules, and norms. Liberal feminists have actively promoted full integration in all positions, branches and ranks, as they see no legal explanations for gendered exclusions and confront gender stereotypes that assume a correlation between masculinity and war, femininity and peace (Devilbiss 1985, 1990; Holm 1982; Segal 1982; Segal and Hansen 1992; Stiehm 1981, 1989). All they argue for is equality in opportunity, and "that the best way to insure women's equal treatment with men is to render them equally vulnerable with men to the political will of the state" (Jones 1984: 75).

While most of the liberal feminist works are concerned with presenting the evidence in favor of integration, very few look at the process of the policy change and its determinants. Judith Hicks Stiehm was among the first and most influential writers to study military policy concerning enlisted women in the United States from 1972 to 1986 and "justifications offered for those policies as they have developed through law, research, and bureaucratic decision making" (1989). Her book is a seminal account of the policy change to demonstrate the military's bias against women and need to implement a fairer policy—for example, resegregation of forces, giving the U.S. Air Force entirely to women, or redesigning equipment to easily fit women. In fact, Stiehm's chapters on draft abolishment, equal rights legislation and legal decisions that played a part in formulating United States military policy regarding women's integration are crucial to our understanding of the role of domestic processes and domestic actors in national security policymaking. Not only did Stiehm write the first book, but she is an example of domestic activists inside and outside of government structures who have been pivotal in opening positions for more women in the military since the 1960s. Stiehm and liberal feminists not only explain the need for gender integration in the military, but, I argue, have become an explanatory variable themselves by framing national legislative debates in terms of gender equality.

The strongest criticism against liberal feminists comes from the *difference* branch of feminist thought, which argues that liberal feminists ignore sexual differences, be it biological, psychological, experience, or socioeconomic conditions. Goldstein puts it: "some difference feminists see such gender differences as biologically based, whereas others see them as entirely cultural, but they agree that gender differences are real, and not all bad" (2001: 41). There are various strands of difference, and scholars, activists and practitioners of a wide variety can be put under this one title but they roughly agree that women are more peaceful and collaborative, while men are more violent and autonomous. But there are divergent thoughts among difference feminists, and these matter particularly as they engage in discussion for and against women's participation in the military. The early difference arguments can be traced back to Virginia Woolf's *Three Guineas* and her argument that:

> "we" ... must differ in some essential respects from "you," whose body, brain and spirit have been so differently trained and are so differently influenced by memory and tradition. Though we see the same world, we see it through different eyes. Any help we can give you must be different from that you can give yourself, and perhaps the value of that help may lie in the fact of that difference. (18)

For Woolf, war is for men, and that it is an expression of "manly" qualities. But Woolf celebrates women's unique attributes and does not view them as a source of inferiority. Rather, she seeks to highlight the special *experience* of woman as the source of positive female values in life and in art. Similarly, many modern difference feminists emphasize the link between peace, pacifism and feminism and view women as more nurturing, cooperative, and less confrontational (Ruddick 1989), and "sees peace politics as resistant to the masculinist and as specifically feminist" (Khanna 2009). Such theoretical explanations are often used to erect legislative barriers to gender integration in the military by suggesting that this type of "difference" will manifest itself in divergent capabilities to fight the enemy among women and men, and by default it would make our militaries less effective. These arguments surfaced immediately after enactment of the All-Volunteer Force that is associated with increased gender integration. According to some, such policies were wasting time and resources on solving "such matters as sexual harassment, pregnancy, joint spouse assignments" (Baucom 1985: 18). Despite the fact that academic and military studies have debunked myths based on women's innate love for peace, genetics, male bonding and unit cohesion (Fraser-Andrews 1991; Saimons 1992; Harrell and Miller 1997; Goldstein 2001), a plethora of authors seeks to find a correlation between gender integration and high divorce rates, adultery, single-parenthood, and high dropout rates to argue against the integration (Tuten 1982; Mitchell 1998; van Creveld 2000, 2001; Gutmann 2000; Fenner and deYoung 2001; Schlafly 2003).

In summation, while feminist security writers reveal centrality of gender in security and hegemonic masculinity in the construction of state's policies on women in the military, such structural explanations treat domestic political actors as agents of change, and international security organizations as sources of gender mainstreaming norms as largely unimportant. On the other hand, liberal and difference feminists offer a glimpse into domestic legislative arguments made to expand and limit gender integration, but consider international security environment as exogenous and have not been systematically comparative. This book seeks to build on both sets of scholarship, and offer a comprehensive theoretical framework for analysis of gender integration in the military.

Previously Proposed Models

Discussion regarding the factors influencing gender integration in the military services is a relatively new one. However, there are four proposed models that have paved the way for future research, including this book. In 1995, building on works such as Stiehm's, Mady Wechsler Segal, an esteemed military sociologist, proclaimed that it was time for a systematic theory of the conditions under which women's role in the military expands and contracts. Her 1995 article "Women's Military Roles Cross-Nationally: Past, Present, and Future" that appeared in *Gender and Society* identifies three categories of variables broadly titled "military, social structure, and culture" and seeks to hypothesize the relationship between each of the three sets of independent variables and her dependent variable, participation of women in the military.

Table 2.1 Mady Segal's proposed model

Military	Social structure	Culture
• National Security situation • Military Technology • Combat to support ratio-Force Structure • Military accession policies	• Demographic patterns • Labor force characteristics • Economic factors • Family structure	• Social construction of gender and family • Social values about gender and family • Public discourse regarding gender • Values regarding ascription and equity

Participation of Women in the Military
(degree of representation and nature of activities)

Source: Segal (1995: 759).

The table presents the variables within each category that are supposed to have some effect on the degree of representation and nature of activities of women in the military, but the way it was presented by Segal, the model does not explicitly establish causal flows. The main argument is that "the military need for personnel has been the driving force behind the expansion of women's military role through history and across nations, but cultural values supporting gender equality also contribute and seem likely to have increased influence in the future." In addition, Segal argues that women are more likely to be participating in modern all-volunteer forces during times of either high or low threats to national security, and serving mostly in administrative and logistical roles. Segal's work is incredibly insightful because she was the first scholar who sought to study the factors affecting women's military roles and who has identified a large number of variables, though her methodology, conceptualization and empirical evidence raise problems. While the author studied integration primarily within NATO member states, different sets of propositions were sometimes applied outside of the sample which makes the methodology questionable. For example, her proposition that "other cultural issues" such as religion and race play a strong role in determining military integration were tested in Israel and South Africa, two obviously extreme cases, and both outside of the original sample. Moreover, there is a serious problem regarding conceptual stretching (Sartori 1970), particularly regarding what Segal terms Cultural Dimension. Most political scientists would agree that conceptualizing and measuring culture is one of the most difficult tasks in comparative politics. According to Segal's empirical evidence, culture is at times synonymous with the "existence of gender discrimination laws," and "religious fundamentalism" among others. While religious fundamentalism can be classified as a cultural variable, the state's gender discrimination policies and laws should by all means be considered institutional or political variables. Despite design and methodology limitations, that Segal acknowledges, one has to applaud her for asking the question, attempting to offer some preliminary answers, and being acutely aware of the importance of the subject as a key to our understanding of civil-military relations.

In 2002, Iskra, Trainor, Segal and Leithauser tried to address some of the shortcomings of Segal's first attempt at theory building. These authors expanded the model by proposing that states with legitimate civilian-led governments, liberal political leadership, egalitarian public policy initiatives, and lack of non-violent sources of social change will have a greater participation of women in the military, testing these propositions in Mexico, Australia and Zimbabwe. Again, the model looks more like a list of possible variables as it lacks explicit causal flows between independent and dependent variables. Yet, these propositions are valuable to this study because they explore domestic political actors and institutions, but their article suffers from the same conceptual and empirical problems as Segal's 1995 article. The authors have adopted the somewhat ambiguous definition of cultural variables that sometimes became synonymous with the existence of anti-discrimination laws, such as the Australian equal opportunities acts of 1984 and 1985, yet clearly these are public policies which they themselves cast as "political variables." Similarly, empirical evidence is problematic in the case of Zimbabwe because propositions were mostly tested against Liberation War guerrilla forces rather than official Zimbabwean Defense Forces. This evidence does not help us find the causes of the state's policy change considering that over 50 percent of women guerrillas were sent home and were not integrated after the Liberation War.

One of the main arguments in the Iskra et al. article is that in societies such as Mexico, cultural values will continue to negatively influence female participation. They particularly argue that machismo will continue to be an obstacle to women, yet only seven years after their article was published, Mexico's elite military academies were open to women, and there was a 61 percent increase in women's applications to those schools in only one year. In addition, what remains unexplored are examples of traditional and patriarchal societies with low levels of threat, such as India, Italy, Libya or Liberia, which have integrated women into their military services. Elsewhere, states such as Eritrea, Malaysia and Peru draft women into the army, while Bolivia, Chile and Guatemala already allow women in combat (unlike the United States). And yet, none of these states has been known as a champion of women's rights. Hence, although their focus on "political variables" brings a much needed discussion of policymakers and institutions within which they operate, the insistence on cultural variables without much agreement on what they are, how they exert influence over policy or to what extent they matter, despite the mounting evidence against it, leaves us without much theoretical explanation of variation.

Additional criticism of these sociological works is offered by Gerhard Kümmel, who argues that all of the previous models neglected international system variables and that, "given the impact of what Robert Keohane and Joseph Nye called complex interdependence, it is deemed appropriate to elevate the international context to a separate category" of variables. Kümmel sought to further Segal's research and sketched out a model that includes the four categories of national level variables (military, social structure, culture and domestic politics), and adds the international environment as a fifth category. He argues that we need to look at the international security circumstances and the way threats change over time, but he has not made any specific hypothesis nor has he tested his proposed model, and has not identified any direct causal flows.

The first actual book to pose a question regarding women's participation in the military is *Gender and the Military*, by Portuguese sociologist Helena Carreiras (2006). While the book was published only in 2006, data presented were from 2000, and hence outdated due to the expansion of NATO and the dramatic changes within both old and new member states' armed forces. The author makes a valid point that the previous works lack parsimony and that concepts need clarification. But the model she develops does not seem to be any simpler. Carreiras opted for a large-N analysis of 18 NATO states that is supposed to eliminate the large number of variables proposed by previous studies.

First, the Index of Gender Inclusiveness (IGI) is developed based on eight different indicators that help her quantify the dependent variable. Next, the author separates her independent variables into two major categories: military "internal" variables (organizational structure, military culture and strategic orientations) and societal "external" variables (socio/economic, political, and cultural factors). While it is presented as two simple categories, in reality these are six very different sets of variables that previous authors have discussed. In addition, the international geo/strategic context and length/history of women's military presence are also considered important factors, and in the model they appear as additional separate dimensions. While previous models have had up to five different sets of categories, Carreiras's "simplified" model includes a total of eight, which unfortunately results in making the model both visually more complicated and methodologically less parsimonious. Despite a large number of variables, Carreiras tests only four independent variables. These are: conscript ratio, the percentage of women in the labor force, Gender Empowerment Measure, and time. The greatest contribution of her study is that time is not a good "predictor of women's representation in the military: a longer presence of women in the ranks does not imply a consistent increase in their relative numbers" (Carreiras 2006: 126). While she offers a great deal of data in her appendix, Carreiras makes a similar "cultural" assumption to previous models. Although her original theoretical model also includes culture, she finds it too difficult to quantify and declares that "culture should be understood as a layer, affecting all other spheres, from the global political and social contexts to the military institution itself" (Carreiras 2006: 21). As a consequence, the book argues that it is simply unnecessary to test cultural variables because "gender values cut across and intersect with the remaining dimensions," but it is unclear how as there is no further exploration of the subject. Instead, the question remains "what is culture?" and how do we specify it, measure it, and then, how does it matter in our analysis of women's military participation. Which particular attitude, cultural trait, norms and practices are associated with higher or lower levels of gender integration in the military is unspecified and does not help in furthering our understanding of the impact that culture has.

Similarly, political variables, social and economic variables are subsumed in the United Nations Development Program's (UNDP) Gender Empowerment Measure and therefore it is impossible to see how they interact. Carreiras proceeds with the qualitative analysis of Portugal and the Netherlands, and content analysis of interviews with female and male soldiers in those two countries to study dynamics of gender integration on the micro-level. Case selection is also problematic because out of 18 cases whose gender inclusiveness index score varies from 0 to 17, Portugal and the Netherlands (11 and 14) do not show much variation. Therefore, although Carreiras attempts to reduce the number of explanatory variables, the model ends up lacking in detail, because it does not include any cultural variables, and it does not differentiate, evaluate and measure individual weight of political, economic and social variables.

The latest addition to the debate comes from Irene Eulriet's book focusing on the uniqueness of the European processes of gender integration in the military, and particularly British, French and German cases. Her research is narrower than any of the previous works, and questions the impact of public cultures of European societies, that despite their liberal orientation toward justice and equality shape asymmetrical policy measures regarding men and women's recruitment in the armed forces. A key argument is that sex distinction results from the internal constitution of the public culture of liberal societies itself. Without assuming cultural homogeneity, Eulriet argues that national differences play a role on both cultural and institutional levels, and that unlike the rest of the previous sociological works focusing on universal explanatory variables, "sex distinction is not universally given and it can only be assessed in situ" (2012: 5). Although Eulriet's understanding of American military sociology is remarkable, her criticism of the abstract models explaining the policy outcomes within the discipline has its limitations. It is correct to suggest that every new version of the model brought new sets of variables, and hence created a long laundry list of possible answers. However, to suggest that it is "hardly workable for empirical purposes" (2012: 14) is hardly a criticism. Although tedious, empirical testing of such models is possible and should have been encouraged by the scholars not only for the sake of parsimony, but as a basic principal of theory generation and falsification. Popper argued that although there is no way to prove that the sun will rise, we can theorize that every day the sun will rise, and if it does not rise on some particular day, the theory will be falsified and will have to be replaced by a different one. In the meantime, there is no need to reject the assumption that the original theory is true. The same can be said for the previous models that Eulriet dismisses without subjecting them to the proper test, and occasionally resorts to agreeing with some of the proposals, but rejects further scrutiny. There is a footnote that reads: "the fact that women will be able to serve on submarines in the British military seems to be related to the U.S. decision to waive its own restrictions in these matters (linked to manpower shortage, apparently)" (Eulriet 2012: 126) in which the author clearly states the importance of manpower shortage but does not follow through with analysis of the statement. Similarly, she points out that although "lots of debates probably took place within the women's movement and elsewhere, but those debates as well as their influence on policy outcomes are difficult to trace through a chosen research strategy" of her book. It begs the question regarding such research design: was it chosen for the right reasons? And ultimately, why was it "chosen," if it ignores possible answers to the research question posed?

While military sociology offers a methodological framework, a great deal of insight, and a strong foundation, I have identified areas that need improvement. In the following section, I extrapolate some of the proposed hypothesis and seek to place them within a larger theoretical discussion on military manpower needs, gender policymaking and gender mainstreaming that can serve as a base for their inclusion in the empirical tests.

Military Manpower Matters: Demographics, Unemployment and Professionalization

All four military sociology models argue that demographic variables, professionalization and unemployment in the civilian sector have an effect on the military recruitment of women. Here, I present the discussion by situating it within security studies literature on military force structure and its modernization in the late twentieth century.

One of the starting points of any discussion of manpower should be the demographics of a state. In modern industrialized societies, there is a need to study the security implications of shrinking youth cohorts due to low birth rates on the further integration of women into the ranks to compensate for the lost recruits. In 1985, Colonel William Hunter, Jr., argued that the American All-Volunteer Force would have to find an alternative to the projected 20 percent decrease in the white male population between 1978 and 1990. According to Hunter, the "American All-Volunteer Force could not have been maintained in the past without a significant increase in the number of women being recruited and without black Americans."[11] The latest study by Rickard Sandell argues that military forces will be the first to feel the impact of the new reality" because "human resources targeted by military recruitment are a population resource in crisis" due to low birth rates.[12] He shows that the armed services, which as an employer rely largely on young men, are starting to adapt to these new realities by moving away from large conscript armies toward small professional military forces that include women. His argument seems to be particularly visible in the developed NATO states. Similarly, RAND Corporation studies argue that the United States needs to pay more attention to demographic variables such as shrinking youth cohorts and low fertility rates in Europe, because they affect the way these developed states will organize their armed services, and meet national security threats in the future (Nichiporuk 2000). Others suggest that besides some highly qualified individuals from the civilian sector, such as doctors, lawyers and scientists, armed services have virtually no "lateral entry" and therefore no way to replace the labor required to sustain large conscript armies (Quester and Gilroy 2002). These works are important to this study because they have already successfully demonstrated the importance of adopting security and personnel policies that reflect the changes in the demographic composition of modern societies.

Previous models mention economic factors and in fact, there is considerable literature that demonstrates that in many states men are simply not joining up as they used to, because there are better economic opportunities in the private sector. Rickard Sandell points out that military recruiters will compete for the same shrinking youth cohorts as both educational and market institutions. Today,

---

11   Hunter (1985).

12   Sandell (2006). Sandell's arguments will be discussed in a larger detail later in Chapter 3 as a key source on the shortage of males in targeted recruitment cohorts (ages 15–30) in the next four decades.

the most popular explanation for recruiting patterns over the last several decades is that the military attracts a lot of young people who are unable to find employment when the civilian job market is in decline and endures a corresponding shortage of labor when job markets are favorable for young people.

In fact, only one study found no correlation between the effects of unemployment and military labor shortages (Ash, Udis, and McNown 1983) and others successfully demonstrated that it suffered from serious data collection and analysis problems (Dale and Gilroy 1985; Brown 1985). Warner, Simon and Payne's research findings show that each 10 percent decline in civilian unemployment results in a 2 percent to 3.5 percent reduction in high-quality enlistments. Similarly, a 10 percent increase in military pay will help increase enlistments by 10 percent.[13] In fact, prior to the September 11, 2001 terrorist attacks in the United States, the Army, Navy and Air Force were encountering trouble recruiting in part due to a booming economy, better-paying jobs in the private sector and a higher number of young people enrolling in colleges.[14] The problem exists elsewhere within the alliance. Keith Harley presents the case of the United Kingdom where the recruitment figures have dropped as well due to low unemployment and the highest number of college students ever.[15] The United Kingdom has already openly stated that the decision to permit women to serve at sea was taken because there were too few men joining the Royal Navy.[16]

Lastly, it is crucial to take a closer look at the modernization and transformation of military forces that have taken place in the years since the fall of the Soviet Union and the end of the Cold War and how they might have affected the integration of women into the services. Throughout the 1990s, states around the world reduced personnel, cut defense budgets, closed their military bases, and formed new volunteer-based forces. As the threat from Communism has disappeared, so have the large mass armies based on conscription that were intended to fight Communist states. Most states realized that as their strategic situation and geopolitical landscape changed, the expenditures of such a military organization were no longer justified (Adams 2008). They moved from a heavy to a light force model, meaning they shifted their emphasis from large, conventional warfare forces, to light, high-quality and high-tech forces designed to conduct asymmetrical warfare (Snow 2007).

---

13   Warner et al. (2001: 41).

14   Sorokin (2002).

15   Harley (2006).

16   "Shortage of men allowed women to join warships," by Michael Evans, *The Times*, July 25 2005 discusses the Memorandum by Admiral Sir Brian Brown, the Second Sea Lord. "Employment of WRNS [Women's Royal Naval Service] personnel in the Royal Navy" written in 1990 but released to *The Times* under Freedom of Information Act in 2005. Available at: http://www.timesonline.co.uk/tol/news/uk/article547749.ece (accessed July 31, 2008).

Besides an institutional change, we have also witnessed the evolution of the art of war, and specifically changes in the operations that ranged from humanitarian interventions and peacekeeping to rapid response, anti-terrorist and counterinsurgency operations. With the new force configuration model, states had to possess a new kind of soldier whose training and skills would be far beyond those of conscripts. In fact, most of the time states did not train their conscripts to use sophisticated technical equipment and they did not send them to conduct missions abroad due to the relatively short duration of compulsory service (Sloan 2001; Williams 2005; Williams and Gilroy 2007). This heavy conscripted-force model was not providing states with the skills that this new era required. Indeed some observers noted that keeping such an expensive and bulky service was a waste of money that could be used instead to update military technology and train professional soldiers (Russell 2003). Now they needed "adaptive, highly-trained, and versatile leaders" (Shadrock 2007: 69). This was particularly the case after September 11, 2001, when it became clear that future security challenges such as international terrorist networks will not be successfully met by conventional military combat forces but rather by smaller, more efficient and more effective forces.

Principal Deputy Under Secretary of Defense for Policy Ryan Henry sums it up by arguing that the new security environment in which we find ourselves today has implications for the kind of manpower we will need. He argues that:

> our sense of the new strategic landscape—and the opportunities opened up by emerging technologies—has led to a new way of measuring military effectiveness. Numbers of troops and weapon platforms are no longer the key metrics. Rather, military effectiveness is now a matter of capabilities—speed, stealth, reach, knowledge, precision and lethality. Thus, our defense planning should place less emphasis on numbers of forward forces than upon capabilities and desired effects that can be achieved rapidly. (Henry 2006: 15)

What followed was the end of conscription in most NATO member states, and it ushered in a new era of professional military. The argument is that states will seek to recruit and retain those who have high-quality professional skills. Thus, the military had to compete in the labor market for the best candidates, while changing their recruitment and compensation system (Moskos and Wood 1988). In order to attract and retain higher-quality soldiers in their new volunteer-based armed services, states started to remove legal limitations to women and invested a great deal of resources in making military services more family and women-friendly places (Shields 1988; Williams 2005). Some authors suggest that integration benefits the armed services because women who enlist tend to be better-quality recruits than men, with more education and better scores on standardized tests (Binkin and Bach 1977; Quester and Gilroy 2002). Therefore, this research will seek to contribute to the literature on military personnel recruitment and labor markets by studying the effect that the shortage of military labor—especially of young educated males—has on the integration and expansion of women's roles in military services.

Domestic F(Actors) Matter: GDP, Women's Presence Inside and Outside Government, and in Labor

Most of the discussion on women in the military has neglected to address and discuss at any length the political actors involved, economic participation and GDP levels that scholars, particularly those within comparative gender policy literature, have identified as crucial in developing successful gender policy. By overemphasizing military manpower needs, security literature has ignored other domestic considerations that help explain women's participation in other non-traditional sectors, such as politics. In her analysis of what makes some democracies women-friendly, Jill Vickers identifies among others high-income levels, a significant "presence" of women in state institutions, including a critical mass of women legislators and programs which empower women (2006). Income also plays a role in Barbara Palmer and Dennis Simon's new book, *Women & Congressional Elections: A Century of Change*, who found that women-friendly districts in the United States also tend to have typically high-income electorates. In fact, today, countries with higher GDP levels have a higher female employment rate and more women in parliament than states with lower levels (2012). One should hence wonder, if that translates into a more female-friendly military.

There are plenty of research works, such as Halsaa's study on the promotion of women candidates in Norway (1998) or Weldon's cross-national analysis of domestic violence policy, that demonstrate the effectiveness of "strategic partnerships" between women in parliament, women's groups, and women's policy agencies (2002). Amy Mazur, for example, identifies women policy agencies as one of the essential government organs that actively promote the advancement of gender equality. In addition, she shows that these "strategic partnerships" and feminist advocacy coalitions have to involve non-feminist allies, such as government ministers, party officials and even the chief executive in order to be successful (2005). Similarly Vickers argues:

> Where women successfully "got in on the ground floor" of new political institutions, moreover, they often did so by piggybacking on other movements. Women's movements must be prepared to consider strategic alliances with other movements, by being open to politics beyond gender issues. In this way, they can develop enough of a "presence" to promote more "women-friendly" states. (2006: 26–27)

A closer study of the integration of women in the United States military in fact reveals the important role that the movement for equal opportunity in labor directly and indirectly played in opening of the military to women (Feinman 2000).

Another branch of gender policymaking literature looks at whether women politicians have a different policy agenda than men. Such research is very similar to feminist "difference" arguments presented earlier. While some claim that there are no legislative differences and that women cannot be treated as a unitary or predictable group (Dolan and Ford 1998), Thomas and Welch in their study of 12 American states demonstrate that "women and men have somewhat different policy priorities. In addition to supporting women's issues to a greater extent than do men, they are now also giving these issues a higher priority than men" (1991: 454). The argument was also tested abroad and the evidence from the Norris and Lovenduski study of the British Parliament seems to point in the same direction (1989). Even more recent studies commissioned by the Center for American Women in Politics support the findings that women legislators are more involved in women's rights and children's issues then men (Carroll 2001). The latest study of women representatives in the U.S. Congress also shows that women are more likely to sponsor and more intensely pursue legislation on women's issues (Swers 2002). In fact, when it comes to the integration of women into the military in the United States, a four-term representative and four-term senator from Maine, Margaret Smith Chase, is often credited with playing a crucial role in the policy outcome. Some argue that without her, "it is certain that momentum would have been lost and the fight would have been bitter and protracted."[17]

Other gender policy scholars, in particular those in the United States, have for many years focused on the argument that the increased representation of women in state and local legislatures makes governments more responsive to women's issues and concerns. What this literature really focuses on is "critical mass"—an idea that "the election of an adequate number of female politicians will result in governance more responsive to women" (Grey 2001: 1).

The concept only arrived in political science via sociology, which in turn borrowed it from nuclear physics, to demonstrate that we need to reach a quantity that will trigger a chain reaction. In political science literature, the chain reaction would be the irreversible process by which legislatures would become more concerned and quicker to respond to women's issues. But what is the actual number of women necessary to reach that tipping point? For some writers, a meager 15 percent has been enough to turn the tide (Grey 2001); others double this (Dahlerup 1988).

However, some authors have recently argued that reaching that magic number is not necessarily going to affect the female legislators' ability to act more and to persuade others to join her in an effort to improve legislation on women. On this side of the debate are scholars such as S. Laurel Weldon, who argues that:

---

17   Sherman (1990: 77). There are many other supporters of this idea including the Margaret Smith Chase Library that claims that she "almost singlehandedly won the permanent status for women in military." Available at: http://www.mcslibrary.org/bio/biolong.htm (accessed October 24, 2008).

the proportion of women in the legislature is likely to have an impact on policy involves a fallacy of aggregation. Such a fallacy occurs when one assumes that something about a whole unit can be understood merely by aggregating the characteristics of the parts. But the whole may be more than sum of its parts ... Most of the research on the impact of women legislators examines individual-level variables, such as policy preferences, legislative or leadership style, and party or other political affiliations. But these differences do not necessarily aggregate in ways that make the activities of legislatures reflect these different characteristics ... Thus, many questions remain as to how the differences that have been documented aggregated in the legislature to affect policy outcomes. (2002: 90)

Weldon's concept of "fallacy of aggregation" seems to hold true particularly outside of the United States where party systems are stronger and legislators have primary loyalty to the party's policy agenda and not to feminist causes (Studlar and McAllister 2002). Studies of laws establishing quotas for women's participation as candidates in national elections in 12 Latin American states also demonstrate that such policies can be passed with even less than 15 percent of women in legislatures (Rodriguez 1998; Htun and Jones 2002). It is therefore possible that previous models have overemphasized the impact that the degree of women's political activity might have on the level of women's participation in the military.

While there is no clear consensus regarding the role that women legislators play in the promotion of women-friendly policies, what all these works underline is the need to take into account the impact that women legislators have on the framing and passing of gender equality policies. There is a danger, however, of assuming that women have an opportunity to participate in the policy process, as that is actually not always the case, particularly in states that are not democratic or that are going through transition. Feminist agendas are often part of the democratization movement but there is evidence that democratization does not necessarily improve women's political influence, and Eastern Europe is a great example: as women's legislative and political participation dramatically dropped, women's legislative quotas were undone, budgets for family programs and childcare cut, and abortion and maternity leave rights threatened (Viterna and Fallon 2008) and their return to the traditional domestic sphere is advocated by new political parties (Einhorn 1993). This book will contribute to this body of research by exploring and empirically testing the impact that female politicians have had on the improvement of women's participation in the military.

Besides studying women inside government, I suggest we study the role that women in the military played in the passage of gender integration legislation. In her influential work on the protest of the feminist movement within the church and military, Mary Katzenstein successfully shows that "feminism in the military engages in an influence-seeking interest-group form of politics" (1998: 18). She argues that the women of the Defense Advisory Committee on Women in the Services (DACOWITS), the Women's Officers Professional Association (WOPA) and the Women Military Aviators (WMA) were integral in keeping equality on the military agenda. In fact, there is no discussion in the previously mentioned theoretical models of different kinds of political activism and lobbying by women's groups regarding the question of military service.

Yet, it is not enough to just explore the general role that women inside government and its institutions play. Rather it is also necessary to explore the effect that women's movements outside of the government have on policy agenda. There are very few works that study cross-nationally the role that women's movements play in policymaking (Gelb 2003). One has to wonder why pressure groups would be left out, particularly if they have been seen as catalysts in the promotion of gender equality in labor. There are no works at all studying comparatively how different types of feminists view the integration of women into the military, or how perceptions of the state and strategies of different feminist movements have an impact on the formulation of policy on women in the military. Single case studies of the United States and Israel demonstrate that liberal autonomous feminist movements have been particularly active in helping women overcome legal obstacles to allow them entry into the military ranks as a central component of equal citizenship, whereas radical feminist movements have seen the integration as just another strategic move by the patriarchal state to co-opt and oppress women (Feinman 2000; Sasson-Levy 2003; Barak-Erez 2007). One of the key issues explored in this book will be the role that such movements play in the promotion of policy of gender integration.

Culture Matters: Religion, Societal Values Regarding Gender in Politics and Economics

All four previous models argue that women's position in society and particularly their presence in non-traditional employment such as politics are explained not only by structural, institutional, or international factors, but also by cultural factors. But as mentioned above, the conceptualization and measurement of "cultural factors" remain rather problematic, ranging from vague to completely irrelevant in all four. But these authors are not the only ones having trouble defining it. One of the biggest problems with studying culture as a dependent variable is the actual conceptualization and measurement. What do we mean by "cultural factors"?

As Harry Eckstein points out, "the term culture, unfortunately, has no precise, settled technical meaning in the social sciences, despite its centrality in them" (1988: 801). Others argue that culture does not differ only from state to state, but from region to region, class to class, or from one ethnic group to another, therefore making it even more difficult to assess and measure as a possible determinant of our national policy outcomes (Narayan 1997).

But that does not mean we should give up on *culture* as an independent variable, but rather that we should try to be as specific in identifying different aspects of it as we can for the sake of both parsimony and theoretical model building (Weldon 2002). Therefore, I will discuss the literature that tests some of the specific cultural indicators as possible determinants of policy outcomes in both the national security arena and gender studies.

Culture has been used to explain why northern European states, such as Norway and Sweden, have higher percentages of women in their parliaments than other Western European states (Karvonen and Per Selle 1995; Solheim 2000). Inglehart and Norris find that in states with what they term "egalitarian culture" regarding women leaders, the proportion of women elected to the lower houses of parliament is higher than in those where traditional values favoring male leadership persist (2003b).

Many political science and, in particular, gender policy scholars have assessed the effect that different types of religions have on gender equality as part of their cultural thesis. Ronald Inglehart and Pippa Norris have been at the forefront of this discussion. They have argued that Catholicism slows down women's progress, and declared that Islam is, however, "the most powerful barrier to the rising tide of gender equality."[18] Elsewhere, they have followed Huntington's "clash of civilizations" argument to contend that democratization alone will not bring gender equality to Muslim states due to the strong influences of Islam.[19] Similarly, others have found that Islam and Hinduism are negatively correlated, while Protestant religions are positively correlated with gender equality (Dollar and Gatti 1999). But as Mala Htun points out, while Anglo-Protestant culture has been found to contribute to the development of liberal democracies and capitalist systems, it does not mean that it has been more supportive of gender equality (2000). On the contrary, she argues, most of the anti-discrimination policies in the United States are so recent that even the Catholic, *macho* states of Latin America, which tend to be hostile to democratic and capitalist ideas, have caught up. Htun's data demonstrate that the status of Latin American women is similar to the status of women in the United States in terms of their political presence, economic participation, wage gap, and existence of anti-discrimination laws. In addition, she shows that although there are "growing structural similarities in the position of women"[20] in both Anglo-Protestant and Latin American cultures, there are still significant differences between the status of women within Latin American states that share cultural traditions and values.

Similarly, there are plenty of examples where women have participated actively in wars, national liberation struggles, and revolutions in the Islamic world. Muslim women have held significant positions in early Islamic history, including the Prophet Mohammad's youngest wife, Aisha, who led troops in the Battle of the Camel in AD 656 (Lindsay 2005). In more modern times, women have not been absent from military actions. Algerian women played a decisive role in undermining French rule (Lazreg 1994). Elsewhere, women have served as military leaders of guerrilla and terrorist groups. Maryam Rajavi is the main leader

---

18   Inglehart and Norris (2003b: 71).

19   Inglehart and Norris (2003a: 62–70).

20   Htun (2000: 193).

of the largest anti-Iranian group, MEK, or People's Mojahedin Organization of Iran (Shahidin 1997). In the Palestinian territories, women were involved in both "military and paramilitary activity within the work of the Palestinian resistance in the 1970s."[21]

Although some would argue that the extent of women's military engagement was largely limited in most of these examples (Shahidin 1997; Taraki 2004), it is important to discuss them because they all demonstrate how some Muslim women have resisted patriarchal values and the use of Islam by more conservative branches. Today women continue to challenge these by joining police forces in Islamic states and official military forces, therefore making it necessary for us to study the effect of Islam on gender inclusiveness.[22]

In addition, to avoid the oversimplification of focusing only on the type of religion as a main cultural variable, this study will also focus on what Norris and Inglehart term strength of religiosity. While almost all of the authors center their research only on studying the effects that a type of religion has on gender equality, Inglehart and Norris, in their book *Rising Tide*, have developed a six-point Scale of Religiosity as a way of measuring actual levels of religiosity in individual states (2003). The following chapter will discuss the six indicators on which the scale is developed in greater detail. This book will adopt their methodology in order to measure the degree to which the strength of people's religious beliefs affects their beliefs regarding gender equality in the military ranks.

International Security Context Matters: International Organizations, "New Warfare" and Threat Level

Scholars have shown that any defense and national security policy analysis must start with the examination of the international framework "within which the state, its bureaucratic agencies and its decision makers are immersed."[23] Similarly, Phillip Cerny urges that the "image of policymaking processes as being constituted predominantly through endogenous variables" is less useful today because it does not account for the change in preferences of domestic actors in a more interdependent world.[24]

As mentioned earlier, this book furthers Gerhard Kümmel's contention that in order to fully understand the changes in gender integration policy, we need to consider international context within which these changes take place. Carreiras also follows his proposal, but her quantitative analysis remains entirely focused on the

---

21   Taraki (2004: 352).

22   The first ever International Islamic Police Women Conference was held in October 2008 in Kabul, Afghanistan, where representatives from Afghanistan, Bangladesh, Indonesia, Iran, Iraq, Malaysia, Pakistan, Tajikistan and Turkey participated to exchange ideas regarding policing.

23   Murray and Viotti (1982: 477).

24   Cerny (2001: 397).

military, socioeconomic and political factors, and there is no empirical testing of international variables. Yet, the relationship between states' integration of women into national militaries and international level variables, particularly international security context within which states find themselves, states' participation in international organizations seeking to mainstream gender and types of missions in the post-Cold War era might be a crucial dimension in our explanation.

When it comes to the relationship between domestic military personnel policy change and international institutions, Gwyn Harries-Jenkins contends that European Court of Justice decisions on the cases of Tanja Kreil *v.* Germany and Angela Maria Sirdar *v.* The British Army Board are crucial determinants of the integration policy in Germany and the United Kingdom (2002). In the former case, the European Court argued that domestic policies that restricted women from "military posts involving the use of arms"[25] were in direct violation of the European Union's regulations regarding gender equality in the workplace while in the latter it only allowed the Royal Marines to maintain their restrictions but it removed all others.[26]

Although not specifically examining women in the military, the edited volume *Service to Country: Personnel Policy and the Transformation of Western Militaries* is one of the chief sources that explore both national choices and international pressures regarding military human resources in the post-Cold War era. The general argument is that while military personnel policies are traditionally affected by national factors such as security threats, demographics and economics, increasingly we are seeing international institutions asserting their role and promoting standardization among member states (Gilroy and Williams 2006). In her chapter, Jolyon Howorth makes a significant contribution by demonstrating that the recent transformation of military personnel policies and military training was due to standardization among European states after the creation of the European Security and Defense Identity and then the European Security and Defense Policy in the late 1990s and early 2000s. The volume also offers a qualitative analysis of the impact NATO membership has had on specific countries, but what remains unexplored is the impact that NATO has had on integration and expansion of the role that women play in the member s tates' military services.[27]

Others would argue that the political and security changes that have taken place since the fall of Communism, including the increase in disaster relief operations, humanitarian interventions, and the less aggressive role of modern militaries, challenged long-held gender stereotypes regarding military participation (Olsson and Tryggestad 2001). As the new internal wars of the 1990s started to feed off state disintegration, it became clear that states are no longer the primary actors in war, but rather "group[s] identified in terms of ethnicity, religion, or tribe" are a new force to be reckoned with (Kaldor 2005: 221). Clausewitz's and traditional

---

25 Raible (2003: 239).

26 Harries-Jenkins (2002: 752).

27 See Urbelis (2006).

conceptions of war with clear front lines, and clear opponents seeking to militarily and strategically outsmart each other in the battlefield, suddenly seemed archaic and obsolete. New actors, unconstrained and often ignorant of international norms, or *jus in bello*, regulating their conduct in wartime, made deliberate use of violence against innocent and immune non-combatants simply as a new way of achieving political goals. It became clear that the very concept of "security" needed to be rethought and adapted to recognize the changing realities of war in the post-Cold War era (Beckman and D'Amico 1995; Sylvester 1994; Zalewski 1995, 1996; Huysmans 1998). These new wars required a different type of response, ranging from traditional peacekeeping, humanitarian interventions to rapid force counter insurgency, and it brought new discussions about the role women should play in these military operations. When it comes to the types of missions in the post-Cold War era, Judith Hicks Stiehm successfully demonstrates that in the five years since the Beijing International Women's Conference there has been an increase in the number of women peacekeepers by at least 3 percent (2001), and others have been looking at the United Nations' attempts to mainstream gender in peacekeeping operations around the world with the passage of Resolution 1325 by the Security Council that mandates Member States to engage women in all aspects of peace-building, including ensuring women's participation on all levels of decision-making on peace and security issues.

Comparative gender policymaking has already identified different international gender equality initiatives regarding family policy, labor inequality, social services, reproductive rights and political representation as causes of domestic policy change (Berkovitch 1999; Riddell-Dixon 2001; Htun and Jones 2002; Naples and Desai 2002; Mazur 2002). This book aims to add to this rich scholarship by examining the impact of international regimes on the integration of women into the military.

Moreover, constructivist literature successfully provides a counterpoint to those in comparative politics who have primarily examined domestic actors and institutions and their impact on domestic policymaking. Some provide a useful example by convincingly demonstrating how international human rights pressures change human rights practices domestically (Keck and Sikkink 1998; Risse-Kappen, Ropp and Sikkink 1999). Finnemore and Sikkink in "International Norm Dynamics and Political Change" successfully outline the evolution of norms through "life cycles," from their international level emergence to their domestic internalization (1998). Equally persuasive is the argument presented by Martha Finnemore, according to which smaller states can be "taught" by international agencies such as UNESCO to adhere to international norms and build national science agencies (1993). Peter Haas, on the other hand, identifies "epistemic communities" as important disseminators of international knowledge, norms and therefore act as a mechanism of social construction (1992). His argument is further supported by Emmanuel Adler, who argues against structural realism and posits that arms control ideas diffused from the U.S. to the USSR, thanks to epistemic communities consisting of MIT, Harvard and RAND scholars (1992). But one important analysis is missing, and that is the impact of international organizations and regimes on the domestic policy of gender integration in the military. Is there such an impact? Have NATO's gender mainstreaming norms trickled down, and how do they interact with the needs and demands of the military establishment, members of the parliamentary or congressional armed forces committees, and women's associational groups? This book answers some of these questions.

Chapter 3

Cross-National Analysis of Gender Integration Policies in NATO: Measuring and Comparing Gender Integration Policies

Comparing the national security policies of different states is always a difficult endeavor and, as previous discussions have shown, very few academic studies have attempted to do so. In order to measure different degrees of integration in the armed forces in different states, this study will first develop a scale, the *Index of Gender Inclusiveness* based on indicators similar to the one first proposed by Segal, Segal and Booth in 1999, and the one developed by Carreiras in 2000 for the sake of theoretical and methodological consistency. Carreiras is rightly concerned with the focus on percentages, as it only identifies the representation of women, but reveals nothing about the diversity of government policies and multifaceted nature of their presence. One way to solve this is to observe and identify areas of policy realignment that promote gender equality in a manner that women's military participation becomes an integral part of military force. This index allows us to employ consistent criteria, or indicators, upon which we can then score the progress each state has made regarding gender integration in the military.

The scale developed in this chapter includes the following six indicators: percentage of total force, occupational restrictions, formal rank restrictions, percentage of women in officer ranks, family programs, and harassment regulations. This scale drops Carreiras's category "percentage of women in traditional functions" and "training segregation." The reason for excluding this category is purely a matter of research design and new legislative realities. Since the writing of her book, the majority of NATO states have, in fact, opened most occupations to women, and thus, "the percentage of women in traditional functions,"[1] becomes irrelevant to the question of government's inclusiveness. The argument here is that unlike eight years ago, it is primarily women themselves and not the legislative and policy changes that are determining the career path. In terms of training segregation, most states have implemented gender neutral training standards, where women and men must meet the same physical standards, and training, instruction and materials, also, remain the same in all. The only exception is Turkey because there are no enlisted women.

---

1 Carreiras focuses on the occupational distribution of military personnel by gender in NATO forces.

State's performance in these six areas is ranked to obtain the final score. The more gender equality policy areas addressed by the state, the higher the score. These policy areas are also chosen as NATO demands their full integration into military practices. In fact, all the data were extrapolated from individual country reports submitted between 2010 and 2012 to the NATO Committee on Gender Perspectives (NCGP) or obtained from individual country delegates on this committee, which is a consultative and advisory body to the Military Committee (MC) on gender related policies for the armed forces of the alliance. Its primary mission is to recommend policy changes, provide advice and guide the military leadership of both NATO Headquarters and individual member states. More specifically, it promotes "gender mainstreaming as a strategy for making women's as well as men's concerns and experiences an integral dimension of the design, implementation, monitoring and evaluation of policies, programs and military operations."[2] It has met once a year since its inception in 1961, and in 1973, a first committee was formed, while in 1976, the Committee on Women in the NATO Forces (CWINF) was officially recognized by the Military Committee. The Office on Women in the NATO Forces (OWINF) was finally established in 1998 in the International Military Staff (IMS) to provide information on gender and diversity issues and support the work of CWINF. Following the passage of the United Nations Security Council Resolutions 1325 and 1820, the committee was renamed NATO Committee on Gender Perspectives and its mandate was expanded to include the incorporation of gender perspectives into NATO operations.

Percentage of Women in Total Active Force Indicator

Most states have, in fact, data available describing the proportion of positions in the armed forces that are occupied by women. While this is the most important indicator, it cannot be studied alone if we are to assess the full scope of integration policy. The number indicates only the quantitative presence, while the other indicators help measure the quality of that presence. The gender inclusiveness index shows that the numbers vary from country to country. The highest percentages of women among the NATO states are: Hungary with 20 percent, Latvia with 17.2 percent, Slovenia with 15.4, the United States and Canada with 14 percent and 15 percent, respectively. At the other end of the scale, states with the lowest percentages are Turkey with less than 1 percent, Poland with 2.1 percent, Italy with 3.9 percent, Denmark with 6.3 and even a relatively low 7.7 percent in Belgium. What these data demonstrate, however, is that the top three states are all new NATO members from Eastern Europe that have only recently opened their ranks to women. Only eight years ago (See Table 3.1), the top three states were the United States (14 percent), Canada (11.4 percent) and France (8.5 percent). Hungary had a meager 6.8 percent

---

2   NATO Committee on Gender Perspectives. Available at: http://www.nato.int/cps/en/natolive/topics_50327.htm (accessed July 2012).

in 2000, but that number has since tripled. In terms of the bottom three, not much has changed. Italy has made the most improvements because it opened its military services to women only in 1999. Therefore, it no longer occupies the last spot on the scale with a zero percentage. Turkey seems to be the one lagging the most, whereas Poland has only recently started to restructure its armed forces, and its numbers are expected to rise in the coming years. It is also clear that all states have experienced an increase in the number of women participating, and seem to be continuing to change their policies to expand even further.

Table 3.1 Percentage of women in NATO Forces 2000 and 2012

Country	Percentage of women in 2000	Percentage of women in 2008–12
Belgium	7.6	7.7
Bulgaria	n/a	14.1
Canada	11.4	15
Czech Republic	3.3	13.8
Denmark	4.2	6.3
France	8.5	14.9
Germany	1.4	8.8
Greece	3.8	5.6*
Hungary	6.8	20
Italy	0	4
Latvia	n/a	17.2
Lithuania	n/a	10.8
Luxembourg	4.2	5
Netherlands	8	9
Norway	3.2	8.6
Poland	0.1	2
Portugal	6.6	13.6
Romania	n/a	4.6
Slovakia	n/a	8.7
Slovenia	n/a	15.4
Spain	5.8	12.2
Turkey	0.1	—
United Kingdom	8.1	9.7
United States	14	15

Source: Data compiled using the annual reports of the NATO Committee on Gender Perspectives.

The majority of the data comes from 2008–12 national reports. Data for Bulgaria, Luxembourg, Portugal and Slovenia are from 2011, while Norway is from 2010 reports. Greek data are from 2007, the last year when Greece explicitly stated that conscripts are not included in the calculation as in other states. For example, after calculating the total number of conscripts, the percentage of women in the Hellenic Armed Forces is 5.6 percent and not 13.7 percent as presented. Since then, the number has exploded to 17.2 percent, though this number is skewed as it does not look at women's presence in the military as a whole, but only among the volunteers.

Occupational Restrictions

This indicator shows whether states ensure equitable opportunity to compete and excel in the armed forces by removing occupational restrictions. Turkey has the highest number of restrictions as it does not allow women to enlist. The only way for women to join the Turkish armed forces is to enter military academies first and then be assigned as officers and non-commissioned officers (NCOs). Many other states have also limited career opportunities for women by excluding them from specific branches or positions. For example, it was only in December 2011 that the United Kingdom government announced that, following an 18-month review conducted by the Royal Navy that looked at the legal, operational, health, social, technical and financial issues of allowing women to serve on submarines, women will have the opportunity to serve on board submarines. The first British female submariners are expected to take up their posts aboard the Vanguard class of Trident vessels at the end of 2013. However, no major changes have been made to the current policy of excluding women from ground close combat roles. The types of post from which women are excluded are those where they may be required to close with and kill the enemy face-to-face, including cap-badged posts in the Royal Marines General Service, the Household Cavalry and Royal Armoured Corps, the Infantry and the Royal Air Force Regiment. In Greece, women are still excluded from submarines, fast patrol boats and Landing Craft Air Cushion-class hovercraft. In the Netherlands, with the exception of the Marine Corps and the submarine service, all posts are open to women. The report cites that these two areas will remain closed to women for reasons of combat effectiveness and practicality, without defining either effectiveness or practicality.

The situation varies among the newer member states in Eastern Europe. In the Czech armed forces, there are no occupations from which women are barred, including combat positions. The 2010 Bulgarian report shows that that there are still specialties that are closed to women in military schools such as mechanized infantry and tank troops, field artillery, fighter pilots, air defense and missile defense troops, and air defense of troops. Poland limits their posts in missile and artillery forces, navy and radio-technical forces due to possible health risks.

Then there are states such as Spain, Portugal, Canada, Norway and Germany that have absolutely no occupational restrictions and women are allowed in combat. However, that does not mean that women are present in all areas. For instance, in the Portuguese army and navy, women can apply for specialties such as Special Operations, Commandos, Marines, Submarines and Combat Divers, but due to very demanding physical selection tests, that are equal for men and women, no woman has yet attained these specialties. Romania has only one exception and that is prohibiting female military personnel from the "religious assistance branch" due to religious practices. In the United States, despite high percentages of women participating, approximately 21 specialties remain closed to women. These specialties are in the following areas: Armor, Infantry, Combat Engineers, Field Artillery, and Special Forces. The debate regarding occupational restrictions continues in the United States, where women are still barred from combat zones, yet on May 1, 2008, Pfc. Monica Brown was awarded the Silver Star, the third-highest combat medal, for her heroism in Afghanistan. She used her body to shield five wounded soldiers from her unit from enemy fire.[3] Sgt. Leigh Ann Hester of the 617th Military Police Company received the Silver Star in Iraq, along with two other men of her unit, for her actions during an ambush on their convoy. She led her team through the "kill zone" and into a flanking position, proceeded to assault a trench line with grenades and M203 grenade-launcher rounds, then cleared two trenches, killing three insurgents with her rifle.[4] The question of women in combat continues to be debated in the United States Congress, where many are calling for an official change to the policy. Just like in many other states, the United States Army has a very clear definition of what constitutes direct combat:

> engaging an enemy with individual or crew-served weapons while being exposed to direct enemy fire, a high probability of direct physical contact with the enemy's personnel and a substantial risk of capture. Direct combat takes place while closing with the enemy by fire, maneuver or shock effect in order to destroy or capture, or while repelling assault by fire, close combat or counterattack.[5]

Such definitions are important because they classify every position based upon the likelihood of engagement in direct combat and are consequently listed as prohibited to women. The Army uses a Direct Combat Probability Coding System (DCPC) to classify every position based upon the likelihood of engaging in direct combat.[6] This issue of occupational restrictions in the United States will be revisited in Chapter 4.

---

3   "Woman Gains Silver Star—And Removal From Combat Case Shows Contradictions of Army Rules," by Ann Scott Tyson, *Washington Post*, Thursday, May 1, 2008, page A01.

4   Sgt. Sara Wood, "Female Receives Silver Star in Iraq." Available at: http://www.army. mil/article/1645/female-soldier-receives-silver-star-in-iraq/ (accessed November 2012).

5   U.S. Department of Defense (1994).

6   For a detailed overview of the coding system see Department of the Army Policy for the Assignment of Female Soldiers, Department of the Army, March 27, 1992. Available

Formal Rank Restriction

Since 2000, almost all NATO members have eliminated restrictions which can prevent women from reaching the top of the military hierarchy. While this analysis will look at the official restriction, it is important to note that just because there are no formal ceilings, that does not mean that women can reach every rank in every branch of the military. Combat restrictions make it hard for women to rise through the ranks, especially in most Army and Marine corps. Many would argue that such rules are contrary to the military's performance-based culture, and are seen as outdated. Jeanne Holm, the first female one-star general in the United States Air Force and the first female two-star general in any service branch of the United States, argued some 30 year ago that, "Any attempt to remove grade ceilings inevitably would lead to pressures to promote women to general and admiral, a prospect regarded in some circles as unthinkable and a threat to national security."[7] Among the last ones to "remove grade ceilings" was Greece, where now both men and women can serve in all positions and have equal opportunities for official promotion and training. The Bulgarian delegation reported in their 2008 survey that there are partial restrictions, but did not offer further details. However, a 2010 report says: "Although in the general provisions on accepting cadets there are no limitations related to application for study of the two sexes, the Services build a 'glass roof' in the career development of women–future military officers because of the fact they are not allowed to be educated in all specialties." One can read this and think that there are *informal* restrictions that exist given the limited access to particular areas of study necessary for rank promotion. The Turkish armed forces seem to be the only ones openly limiting promotional opportunities, primarily because women are not allowed to enlist as privates, and therefore cannot attain ranks that demand field experience.

Percentage of Women Officers

While most states have opened military academies to both men and women and eliminated all rank restrictions, the actual percentages of women who have successfully climbed the ladder varies from state to state. Due to the recent modification of the military rank structure to synchronize with the NATO standards, some states have eliminated ranks and others have integrated new ranks. However, the biggest reasons for variation stem from the exclusion of women from combat occupations, and the timing of the changes in the states' legislation regarding the opening of military academies and specialized occupations. In fact, a large number of national reports argue that women will be in commanding posts in the future,

---

at: http://www.apd.army.mil/pdffiles/r600_13.pdf (accessed April 17, 2012).

   7   Holm (1982: 195).

because it takes time to gain the necessary education, field training and experience to be promoted to a higher rank.

Table 3.2 Percentages of women by rank

Country	Percentage of women officers	Percentage of women NCOs	Percentage of women enlisted
Belgium	7.44	6.6	10.4
Bulgaria	2.4	6.98	15.67
Canada	15.4	12	16
Czech Republic	1.98	N/A	N/A
Denmark	6.5	3.9	6.7
France	9.8	13.5	14.2
Germany	6.98	5.86	10.72
Greece*	N/A	15.9	N/A
Hungary	19	29	11
Italy	1	N/A	2.6
Latvia	12	24.1	13.4
Lithuania	10.3	10.5	13.5
Luxembourg	4.91	3.8	7.28
Netherlands	8	7	11
Norway	7.7	11	8.5
Poland	2.4	0.8	0.5
Portugal	11.6	6.64	17.84
Romania	5.29	5.46	0.47
Slovakia	4		
Slovenia	19.71	10.2	15.4
Spain	5.61	1.17	18
Turkey	N/A	N/A	0
United Kingdom	11.9	8.28	9.5
United States	15.78	13.8	17.9

Note: *Greece does not offer cumulative percentages (3.1 percent senior officers, and 10.61 percent junior officers).
Source: Data used for officers are from 2008 reports of the Committee on Women in the NATO (now NATO Committee on Gender Perspectives) and surveys completed by country delegates in 2009.

The majority of the state reports, however, do not indicate a time period within which they expect equal representation of women and men in higher ranks. To speed up the process, many states have sought to shorten the length for rank promotion so that women who take maternity leave would not lag behind male colleagues. Table 3.2 above indicates the percentages of women officers, women NCOs, and enlisted in all 24 states studied. The states with the most women officers are Hungary (19 percent), Slovenia (19 percent), the United States (15.78 percent) and Canada (15.4 percent). The states with the fewest female officers are Italy (1 percent), the Czech Republic (1.98 percent) and Bulgaria (2.4 percent). However, it must be kept in mind that Italy only opened its ranks and academies to women less than a decade ago. Time can therefore explain the relative shortage of women officers, but if the numbers continue to remain low for another decade, clearly we will need to look for different explanations. In addition, data regarding the Turkish and Greek armed forces remain a mystery. All of the data for both states indicate the actual numbers of women in officer ranks rather than percentages of total force. Such numbers obscure the degree of inclusiveness and do not allow for a closer study of gender balance.

Family Programs

In order to measure how inclusive each one of the 24 governments has been, it is crucial to assess the scope and breadth of the programs that address work—life balance. Therefore, here I have compiled data regarding maternity and parental leave in each state, the existence of child care provisions and programs, family assistance, and flexibility of hours.

The specific family programs for military personnel vary dramatically from state to state. Most of the states claim to have adopted civilian labor codes and regulations regarding family policy. However, there are quite a few that have gone beyond that and implemented specific policies addressing the needs of military families. For example, in Germany, an "operational test" is intended to be conducted over the next two years which would allow both male and female soldiers to bring their children to work in case of emergencies. The German report states that "the aim is to reduce the stress, for example, linked with having to make emergency arrangements or for looking for alternative care" (2008: 4). In Belgium, military personnel have a choice of starting their workday any time between 7 a.m. and 10 a.m. and end between 3 p.m. and 6 p.m. so that they can take their children to school and collect them afterward. In terms of maternity leave, some states are more generous than others. Greece leads with 56 weeks of paid maternity leave, while the United States is the last with only 42 days on average of convalescent leave. Slovenia offers maternity leave and post-maternity rights to all members of its force regardless of gender. Others have focused on policies regarding the work of pregnant women and mothers of young children. For example, in the Netherlands women with children under the age of 5 are not obliged to be deployed. In Romania, women who choose to return to work before a child turns one are able to work only six hours a day while receiving full salary. And in Spain, the opening of 16 nursery schools for children of military personnel has already resulted in an increase in the female recruitment rate, according to its report.

All in all, across NATO states have started to dramatically expand their family programs by providing child care facilities, extend maternity leave rights to both parents, and create more flexible working hours. Not all have done it to the same extent, so we can see some variation in Table 3.3 below. The Czech armed forces have no special provisions or policies regarding family life yet. According to the Czech delegate to the NATO Committee on Gender Perspectives, Magdalena Dvorakova, "CZAF is following the civil law and doesn't have a nursery system for females and other institutions to help raising the kids of single-parent. Mostly, we depend on family support."[8] In fact, social and family policies have been cited in several country reports as one of the biggest reasons why the percentages of women remain low. Closer inspection of the index below reveals exactly that—Turkey, Italy, Denmark, the Czech Republic and even France all have fewer social and family policies, and lower percentages of women.

Harassment Regulations

Many scholars, feminists and activists would argue that it is much harder to be a woman soldier than her male counterpart because at times an enemy is hiding within one's own ranks. Researching the subject and trying to assess both the quantity and quality of the policies that states have passed to prevent sexual harassment in the military is a challenging task. While policies might have been passed by the state's legislative body, and monies allocated to run the workshops, print the brochures and build the counseling centers, it is common knowledge that sexual discrimination and harassment persist. The true number of sexual assaults on female soldiers is hard to ascertain, in part because crimes frequently go unreported.

Recent analysis of sexual harassment in the U.S. armed forces showed how widespread the problem remains, and how sexist behavior, intimidation and rape still occur in the daily life of a female soldier. In fact, the numbers are incredibly disturbing. The *New York Times* reports that:

> new data released by the Pentagon showed an almost 9 percent increase in the number of sexual assaults reported in the last fiscal year—2,923—and a 25 percent increase in such assaults reported by women serving in Iraq and Afghanistan ... The truly chilling fact is that, as the Pentagon readily admits,

---

8   Personal correspondence with author, October 16, 2008.

the overwhelming majority of rapes that occur in the military go unreported, perhaps as many as 80 percent. And most of the men accused of attacking women receive little or no punishment.[9]

The problem is not unique to the United States. In November 2000, Belgian Army female military personnel gathered in Brussels to discuss the problems. About 92 percent of women reported that they have experienced sexist talk, 33 percent of women suffered physical harassment, 33 percent received continuing sexual propositions, and 25 percent have been psychologically terrorized.[10] Does this mean that Belgium has not tried to protect its female soldiers? Since 1997, the Belgian government has tried to solve the problem by establishing a confidentiality unit, but Defense Minister Andre Flahaut himself admitted that "64% of the personnel have never heard about the unit."[11] In Eastern Europe, the Bulgarian government has been under a lot of pressure following several scandals, including the harassment case against Colonel Alexander Petkov for sexually harassing five female soldiers in his brigade.[12]

Yet none of this means that the United States, Belgium, Bulgaria or any other state that is encountering sexual harassment problems in its military service has not done anything to prevent it. Of all states included in this study only the Hungarian and Czech military delegates to the NATO Committee on Gender Perspectives reported not having any specific policies regarding sexual harassment in the military, but rather they use civilian labor laws and their democratic constitutions as guarantees of protection for all citizens regardless of gender. Most states have not only passed policies but also ask that soldiers undergo training specifically addressing respect for diversity, appropriate attitudes regarding sexual harassment, mobbing, or any other kind of violent and disrespectful behavior. All members of the Belgian Royal Armed Forces can undertake three-day diversity training, including gender education.

German forces have only recently opened all positions to women, but they have also been the most active in implementing gender mainstreaming requirements at all schools of the armed forces. Soldiers study topics such as "men and women in the Bundeswehr," "discrimination, mobbing, sexual harassment," "partner-like behavior," "compatibility of family and work" and "communicative behavior" to

---

9   Bob Herbert (2009).

10   Data and charts are available on the RoSa- factsheet regarding women in military whose aim is to familiarize general public with the scope of equal opportunities in Flanders. Available at: http://www.rosadoc.be/site/maineng/pdf/07.PDF (accessed April 14, 2009).

11   Original quote appeared in "25 Jaar vrouwen in het leger: een balans," in De Draad van Ariadne, Number 14, October 2000 and cited by RoSa-factsheet report. Available at: http://www.rosadoc.be/site/maineng/pdf/07.PDF (accessed April 14, 2009).

12   "Bulgarian Colonel Appeals Sentence for Sex Abuse." March 14, 2007, *Sofia News Agency* online. Available at: http://www.novinite.com/view_news.php?id=77880 (accessed March 18, 2008).

eliminate inappropriate conduct between sexes. Greece, on the other hand, has only started to run its gender awareness education pilot program in its air force academy with the hope of extending it to the army and navy.

It is necessary to remember that states will only be scored based on the existence of policy dealing with the question of harassment and not the question of policy effectiveness. Canada, the United States, the United Kingdom, Norway, the Netherlands, Belgium and Denmark all integrated women earlier and therefore have been trying to solve problems relating to harassment for many years. However, time or money invested do not seem to make any of these states more immune to the problem than other states that are just starting to incorporate diversity programs and workshops into their military school curriculums. Table 3.3 below shows how different indicators were measured in my index of gender inclusiveness and how each state was scored. It is this index score that will be then tested against the possible explanatory variables in the following section.

Empirical Findings

Empirical testing and findings are divided into four sections reflecting the four analytical categories of theoretical model presented by previous models: military manpower needs, domestic political and economic factors, culture and international security context. In each section, I tested a set of arguments quantitatively using data from all 24 NATO states. In the first section, I explored whether greater gender integration takes place due to the abolition of conscription, negative birth rates, high percentages of women in professional and technical fields and low unemployment rates. In the second, I looked at gender equality in both the politics and economics of the state to see if there is a spillover from the civilian sector into the military sector. Whether cultural factors such as type of religion and levels of religiosity or societal values regarding the appropriate role for women have an impact on the levels of inclusiveness in the military is the focus of the third section. And in the last section, I examined the effects of international security context and membership in NATO on domestic military personnel policy formulation.

Table 3.3 Index of gender inclusiveness

Country	Percentage of women in total active force (1)	Occupational restrictions (2)	Formal rank restrictions (3)	Percentage of women in officer ranks (4)	Family programs (5)	Harassment regulations (6)	Index score
Belgium	2	3	2	2	2	2	13
Bulgaria	3	2	1	1	2	2	11
Canada	4	3	2	4	2	2	17
Czech Republic	3	3	2	1	1	0	10
Denmark	2	3	2	2	1	2	12
France	3	2	2	2	1	1	11
Germany	2	3	2	2	2	2	13
Greece	2	2	1	0	2	2	9
Hungary	4	3	2	4	2	0	15
Italy	1	3	2	0	1	1	8
Latvia	4	3	2	3	2	1	15
Lithuania	3	3	2	3	2	1	14
Luxembourg	2	3	2	1	2	2	12
Netherlands	2	2	2	2	2	2	12

Notes:

1. Percentage of women in total active force: 0 = 0–2%; 1 = +2–5%; 2 = +56–10%; 3 = +11–15%; 4 = +15–20%; 5 = 21+%

2. Occupational restrictions: 0 = total (no women at all); 1 = many (no enlisted women); 2 = few (submarines, Special Forces); 3 = none.

3. Formal rank restrictions: 0 = total; 1 = partial; 2 = none.

4. Total percentage of women officers: 0 = 0–2%; 1 = +2–5%; 2 = +5–10%; 3 = +11–15%; 4 = +15+%.

5. Family programs (maternity programs, child care, paid leave): 0 = none; 1 = few; 2 = many.

6. Harassment regulations (anti-discrimination regulations and monitoring within the armed services): 0 = none; 1 = few; 2 = many.

Country	Percentage of women in total active force (1)	Occupational restrictions (2)	Formal rank restrictions (3)	Percentage of women in officer ranks (4)	Family programs (5)	Harassment regulations (6)	Index score
Norway	2	3	2	2	2	1	12
Poland	0	3	2	1	1	1	8
Portugal	3	2	2	3	2	2	14
Romania	1	3	2	2	1	1	10
Slovakia	2	3	2	1	1	1	10
Slovenia	4	2	2	4	2	1	15
Spain	3	3	2	2	2	2	14
Turkey	0	1	0	0	1	1	3
United Kingdom	2	2	2	3	2	2	13
United States	4	2	2	4	2	2	16

My initial arguments presented in Chapter 1 are confirmed by the statistical evidence, thereby eliminating a large number of previously proposed determinants of gender integration. First, the states that are still conducting active conscription, and keeping large mass armies, continue to be the least responsive to gender integration into the military while the states that have undergone recent professionalization and modernization are more responsive. In addition, the evidence validates the argument that states with a high degree of gender equality in the civilian economic sector, as measured by higher levels of economic activity and percentages of women in technical and professional positions, are more likely to expand integration in the military. Lastly, unlike all the previous models, this study demonstrates that cultural values are no longer main determinants of integration. With the exception of a correlation between greater gender inclusiveness and the existence of autonomous women's movements, none of the other domestic political, economic or cultural variables seem to have an effect. The analysis shows that states' accession to NATO affects the timing of the policy change, and for the most part this seems to be more pronounced in Eastern Europe. One of the most important findings is that there are differences in the reasons why the original and new members of NATO pass gender integration in the military policies. In the original states in Western Europe and North America, the existence of active women's movements pushing for equal opportunity in all areas show a very high correlation to a higher degree of integration in the military, so does the gender equality in the economic sector. In the new member states in Eastern Europe, the main predictors are NATO membership itself and military operational capabilities.

Military Manpower

The Greater the Shortage of Men Due to Abolition of Conscription, the Greater the Gender Inclusiveness

Since the end of the Cold War, most states within NATO have changed their personnel policies, particularly regarding the compulsory service of young men and general reduction of numbers in active and reserve forces. With the exception of Turkey, Greece, Norway, Denmark, Germany and Lithuania, all NATO member states have officially abandoned conscription and fully adopted a new all-volunteer and professional armed services model.

Table 3.4 Status of conscription in NATO states

Country	Status of conscription	Type of force structure
Bulgaria	ended in 2008	AVF
Canada	Never had conscription	AVF
Czech Republic	ended in 2005	AVF
Denmark	keeps conscription	Pseudo-conscript force
France	ended in 2001	AVF
Germany	keeps conscription	Pseudo-conscript force
Greece	keeps conscription	Hard core conscript force
Hungary	ended in 2004	AVF
Italy	ended in 2005	AVF
Latvia	ended in 2006	AVF
Lithuania	keeps conscription	Pseudo-conscript force
Luxembourg	ended in 1967	AVF
Netherlands	ended in 1996	AVF
Norway	keeps conscription	Soft core conscript force
Poland	phasing out started in 2008	Soft core conscript force
Portugal	ended in 2003	AVF
Romania	ended in 2007	AVF
Slovakia	ended in 2006	AVF
Slovenia	ended in 2004	AVF
Spain	ended in 2001	AVF
Turkey	keeps conscription	Hard core conscript force
United Kingdom	ended in 1963	AVF
United States	ended in 1973	AVF

Source: Data compiled using the annual reports of the NATO Committee on Gender Perspectives.

The reasons behind the elimination of a compulsory military service vary from purely economic consideration and the transforming international security environment, to changing demographics and dramatic increases in conscientious objectors.[13] Some states have already completed the transition, while others are

---

13   For a further discussion of possible reasons see Christopher Jehn and Zachary Selden, "The End of Conscription in Europe?" National Security Division, Congressional

slowly phasing out their compulsory military service, and yet some are choosing to keep their mass military model. Based on data collected from various sources,[14] Table 3.4 demonstrates some of the most recent changes that have taken place in individual member states.

To make the measurement possible, I have divided states into four categories following the typology of force structure by Karl Haltiner. He divides states into four categories depending on the conscript ratio, which is "defined as the percentage of conscripts compared to the total of country's regulars (not counting reserves)."[15] These categories are: All-Volunteer Force (with zero conscripts), Pseudo-Conscript Force (conscript ratio below 50 percent), Soft Core Conscript Force (conscript ratio between 50 percent and 60 percent) and Hard Core Conscript Force (conscript ratio above 60 percent).

The model shows a strong, negative and significant correlation between the index of gender inclusiveness score and conscript ratios ($R = -.587$, $p = 0.010$). This negative correlation suggests that when the level of conscripted military personnel is higher, then the level of gender inclusiveness is lower. It confirms the hypothesis that the greater the shortage of men due to the abolition of conscription, the greater gender inclusiveness in the military. In addition, the model reveals another interesting relationship between the year that conscription was abolished and the year that women were admitted ($R = 0.549$, $p = 0.015$). Greece, Norway, and Turkey still have not announced possible dates for the abolition of conscription and have therefore been excluded

The Lower the Birth Rates in the State and Ratio of Men to Women, the Greater the Gender Inclusiveness

All previous models have argued that birth rates will affect the degree of gender inclusiveness in the military. Moreover, many state reports to the NATO Committee on Gender Perspectives cite demographic changes as one of the main reasons for the inclusion of women and expansion of their role in the military. While it is true that almost half of all NATO member states have had negative population growth rates in 2009, the other half is still experiencing positive growth. The result does not show any significant correlation to my dependent variable ($R = -.152$, $p = 0.477$). This is contrary to the proposed hypothesis that the gender inclusiveness score is determined by the population growth rate.

The data show that a negative trend seems to be more pronounced in Eastern Europe, and that Western Europe and North America population growth rates

---

Budget Office United States Congress, 2000. Available at: http://www.rand.org/pubs/ monographs/MG265/images/webS0228.pdf (accessed March 27, 2009).

14   I have used the International Institute for Strategic Studies *Military Balance 2006–2007* and have updated their data based on the information available on individual states' Defense Department websites.

15   Haltiner (2006: 366).

remain positive in all states, except Germany and Italy. Yet there is no observable relationship between the IGI scores and population growth. States with the lowest scores, such as Poland and Italy, have low population growth, but Turkey has the highest. Similarly, while both the United States and Hungary have high IGI scores, the former has positive while the latter has negative population growth rates. Clearly the only observable pattern is a rather chaotic one, and there is no evidence to support the idea that lower population growth rates are associated with the degree of women's integration into the military.

The Higher the Percentage of Women in Technical and Professional Fields, the Greater the Gender Inclusiveness

The result confirms that there is a weak positive correlation between the percentages of women in technical and professional fields in civilian society ($R = 0.482$, $p = 0.020$) and IGI. In addition, this variable alone explains 27 percent of the variation between the IGI scores (R squared $= 0.269$). States with the lower percentage of women in professional and technical fields in civilian society tend to have lower IGI scores, while the states with higher percentages have higher IGI scores. Although we should be able to argue with confidence that this variable is a good predictor of gender integration in the military, a visual inspection of the scatterplot shows that Turkey is skewing the result. Secondly, the visual presentation reveals another rather interesting point. All of the new member states in Eastern Europe have well over 50 percent of women in professional and technical fields, while most of the original membership is below that number with only three exceptions: Denmark, United States and Canada. I will explore differences between new and old members in a greater detail later in the chapter.

The Lower the Unemployment Rates, the Greater the Gender Inclusiveness

The hypothesis that the lower the unemployment rates the greater the participation of women in the military seems not to be supported as there is no direct relationship between the two ($R= -.282$, $p = 0.182$). States such as Turkey and Spain have relatively high levels of unemployment, yet Turkey has the lowest and Spain is among the higher IGI scores. Similarly, Italy and Canada both have slightly more than 6 percent unemployment, but when it comes to the IGI, Italy is at the bottom and Canada on top of the list. This means that governments do not seek to integrate gender into their military ranks during times of low unemployment. The evidence clearly reveals the lack of an observable pattern and therefore lack of explanatory power of this variable.

Domestic Political and Economic Context

The Greater the Percentage of Women in Legislatures, the Greater the Gender Inclusiveness

Most authors have argued that if we have a higher number of women in politics, we will have a higher number of women-friendly policies. Contrary to this contention, I noted in Chapter 1 that so far there has not been much evidence that women's mere presence in a state's legislature will have any impact on the degree of gender integration into the military. In fact, the numbers show no relationship between the percentages of women in legislatures and IGI (R = 0. 239, p = 0.262). There is no clustering of states and no pattern to support the argument that more women in politics is going to lead to more gender inclusiveness in the military. For example, the United States, with the highest IGI scores, has a much lower percentage of women (16.8 percent) while Italy and Poland, states with the lowest IGI scores, have 21.3 percent and 20.2 percent, respectively. Even more surprising is Hungary, which has the highest percentage of female soldiers, 20 percent, but only 11.1 percent of female legislators. Therefore, this variable has no explanatory or predictive power.

The Greater the Percentage of Women in Ministerial Positions, the Greater the Gender Inclusiveness

As discussed in Chapter 1, many scholars have argued that it is not the quantity but the quality of women's legislative positions that matters when it comes to presenting and supporting legislature dealing with gender equality. I find no correlation to support this argument. Once again, there is only a chaotic pattern. While Slovenia and Spain share similar IGI scores, Slovenia has 18 percent while Spain has 44 percent of women at ministerial level. The United States and Italy have the same 24 percent, yet the former has the highest and the latter the lowest IGI scores among the original NATO members. In fact, the correlation confirms that there is no relationship between the two with R = 0.240, and p = 0.258.

The Greater the Percentage of Women in the Labor Force, the Greater the Gender Inclusiveness

Next among the variables tested in this section is the labor participation of women (percentage of population aged 15–64). There is a very strong and significant positive correlation between female activity in the labor market and IGI (R = 0.743, p = 0.000). What is interesting about this specific result is that it is completely contrary to the result that Carreiras's model produced in 2006.[16] She tested the

---

16   Carreiras (2006: 125).

female economic activity against her IGI scores. It is important to note that I have opted for testing labor participation of women (as calculated by the World Bank and OECD) as opposed to female economic activity (as calculated by the ILO and processed by the UNDP Human Development Index) that Carreiras has used.

Because Turkey is dramatically skewing the result, I repeated the model once again without it and still found a very significant and moderately strong positive correlation (R = 0.551, p = 0.006). Therefore, it is safe to confirm the validity of the argument that the level of gender inclusiveness in the military will be higher in states with higher levels of female economic activity. The graph shows that the United States and Canada, with the highest percentages of working women, have the highest IGI scores, while the opposite is true for states such as Italy, Greece and Poland, which have the lowest percentages of working women. When Turkey is kept as a part of the model this variable also alone explains 55 percent of the variation (R squared = 0.552), making it an excellent predictor of how well women will be integrated into the military. On the other hand, when Turkey is taken out that number drops to about 30 percent (R squared = 0.304). Although the percentage is much lower, this single variable remains a strong predictor of the level of gender inclusiveness.

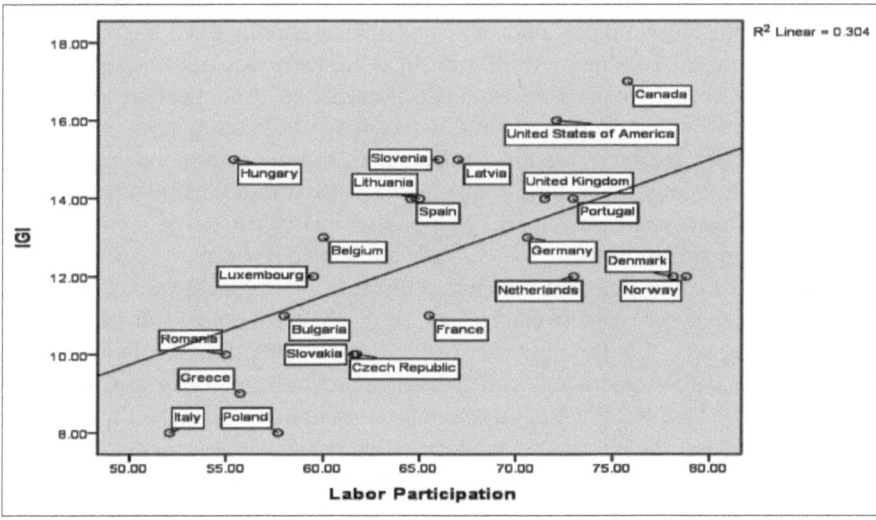

Figure 3.1 IGI and labor participation—without Turkey

*The More Economically Developed the State, the Greater the
Gender Inclusiveness*

There seems to be no correlation between the level of a state's economic development measured by the Power Purchasing Parity GNI per capita and IGI score (R = 0.284, p = 0.178). There is a significant variation in the IGI scores among some of the poorest states with a GNI below $20,000. Latvia, Hungary and Lithuania have among the highest IGI scores, while Poland and Turkey have the lowest. Similarly, there is a variation among the richer states with a GNI between $30,000 and $60,000. The data demonstrate that although Eastern European states are still less developed than Western European and North American states, their IGI scores are not lagging behind. Therefore, there is not enough evidence to argue that this variable will be a very good predictor of the IGI.

*The More Autonomous Women's Associational Groups Involved in Defense
Policy, the Greater the Gender Inclusiveness*

Unlike previous models, I have argued that we need to study the impact of women's organizations, as they tend to be the main conduit able to articulate women's issues and problems to parliaments and parties. More specifically, I have hypothesized that states with autonomous women's movements should have a greater IGI score. Including all 24 states in this statistical analysis was not possible. First, it would be incorrect to include women's movements from Eastern European states because for the most part of their history they were acting as an arm of the Communist Party and therefore did not have a choice to become independent of government structures. Therefore, I am studying only those movements that had a choice between working as an arm of a political party and being a completely independent organization with its own goals and agendas. Second, I am studying the movement at the time that the policy discussion was taking place and not today. The movements tend to change and so do their demands. But in order to account for the role that the movement played in the integration it is necessary to consider the way it articulated its demands at the time that policy was passed.

I have created a dummy variable where the autonomous movement was coded (1) and the non-autonomous movement—or working within existing political parties and caucuses—was coded (0). The findings suggest that there is a very strong, positive, and significant relationship between the two variables (R = 0.689, p = 0.005). It tells us that the states with autonomous women's movements have higher IGI scores. Moreover, it explains 32 percent of the variation in my dependent variable.

The other interesting statistical finding is that it is also positively, significantly, and strongly related to percentages of women in the military (R = 0.781, p = 0.001). States with autonomous movements tend to have more women in their ranks. In fact, Italy, Turkey, Greece, Denmark and Luxembourg all have relatively low levels of gender integration in the military and have all had non-autonomous movements at the time the policy discussion was taking place.

Culture

The Less Religious the Population, the Greater the Gender Inclusiveness: The More Protestant the Population, the Greater the Gender Inclusiveness

In terms of the cultural dimension, of three religions, only Islam is showing a significant and strong negative correlation (R = −.617, p = 0.011). However, I will not claim that there is a pattern or that this variable has a strong explanatory power because it is based on a single observation—Turkey. The results also suggest that there is no correlation between Catholic (R = 0.139, p = 0.517) and Protestant religions (R = 0.283, p = 0.180) and IGI. This is contrary to the literature and the original hypothesis that expected Protestant states to perform better than Catholic states. Among Catholic states are some of the best performers (Spain, Portugal and Slovenia) and some of the worst (Poland and Italy). Therefore, it would be wrong to suggest that Protestant states will be more accommodating than Catholic ones.

Some have been critical of this argument alone and have suggested that it is necessary to test for levels of religiosity.[17] It is common knowledge that people in states such as Italy and Poland are more devoutly Catholic on average than people in other states. In Chapter 2, I have discussed the adoption of the Inglehart and Norris Strength of Religiosity Scale and it shows that the United States and Canada, the states with the highest IGI scores, have relatively the same levels of religiosity as Turkey, Poland and Italy, the states with the lowest IGI scores. The results suggest that there is no correlation between the Strength of Religiosity and my dependent variable (R = −.298, p = 0.157) and thus the hypothesis is not valid. The extent of women's integration into the military services is not affected by the level of religiosity of the population.

The More Egalitarian the Values Regarding Women in Business, the Greater the Gender Inclusiveness: The More Egalitarian the Values Regarding Women Politicians, the Greater the Gender Inclusiveness

The model included variables that tell us whether the way the society in general views women as fit to fill positions in politics and business has any effect on the IGI score. The results are mixed. There is a very strong, negative, and significant relationship between IGI scores and the way that the society views women as capable of filling executive positions in business (R= −.719, p = 0.006). This result would allow me to argue that societies where the majority of people agree that

---

17   I thank Dr. Irving Leonard Markovitz and Dr. Joan Tronto for offering feedback and urging me to add "levels of religiosity" to my model.

men are better business executives will, in fact, have lower gender inclusiveness in their armed forces. However, before such a statement is made it is important to acknowledge the limitations of this correlation. What remains problematic in this case is the actual number of observations and Turkey. Unlike most of the hypotheses in this study, this one was tested only in 13 cases, and not 24. Unfortunately, this World Values Survey was not conducted in all of my sample states, and therefore it is possible that this smaller sample is skewing the final result and producing a rather large correlation. In fact, once Turkey is excluded from the model, the correlation does not reach significance level in this small sample ($R = -.469$, $p = 0.124$). Therefore, the evidence is rather inconclusive and a more in-depth study of the relationship should be conducted once all the information is available.

When it comes to public opinion on whether men are better politicians than women, the result shows no correlation to the IGI score ($R = -.276$, $p = 0.252$). This test was conducted using data on 13 states from the 2005–2008 World Values Survey wave plus data from an additional six states from the 1999–2000 wave. The results show that although the public in Italy and Canada share the same views, they are at opposite ends when it comes to the IGI scores. And even though more than 50 percent of the public believes that men are better politicians in Hungary, Lithuania and Latvia, those countries still have among the most highly integrated military services. Hence, we can argue with confidence that the degree to which women are integrated into the armed forces is not affected by societal views regarding women as political leaders.

International Security Context

The More Defensive Military Capabilities, the Greater the Gender Inclusiveness

Previous models have argued that gender integration is conditioned by levels of threat. While there were vague assumptions about what constitutes a threat and how we identify it, none of the models dedicated much time to the exploration of the concept itself. The reason for this is simple. It is difficult, and it depends on who is the researcher asking the question as I have discussed in Chapter 2. How we define a threat is conditioned upon our definition of security, and we have realist, critical, and feminist interpretations of these concepts. Is it posed by humans, economic instability or environmental changes? Is it sporadic or constant? What is the likelihood of the threat actually becoming a violent event? And how do we account for different definitions of a threat in different states due to their political and socioeconomic peculiarities? Asking all of these questions and building a matrix to account for and incorporate all possibilities is a research project on its own.

In this study, I have sought to look beyond the concept itself, and measured states' reaction to the threat that might be posed to each one of them. That way we can account for those definitional differences, as well as go beyond biased perceptions of threat and still be able to generalize. In order to do that, I have used the Conflict Intensity Scale developed by European researchers Julian Lindley-French and Franco Algieri in the Venusberg Report that ranks states according to their military's operational capabilities and therefore their ability to face a threat. All the states in my study, with the exception of Luxembourg, Norway and Canada, have been ranked and the scale shows how different states are with "defense diplomacy at one end ... to robust preventive missions, possibly anywhere in the world"[18] on the other. Contrary to the previous findings, the data show no correlation between my dependent variable IGI and a state's military preparedness and capability to fight potential enemies ($R = -.036$, $p = 0.879$).

However, the model reveals another very important, significant, and strong negative correlation between the timing of women's integration and level of military preparedness ($R = -.547$, $p = 0.010$). It demonstrates that states with high levels of military preparedness are Western European and North American states that seem to have admitted women into their ranks earlier than their Eastern European counterparts. Nevertheless, that does not mean they have done more than pass the legislation, because the level of military preparedness still fails to predict both the percentage of women in the ranks and the IGI score.

The Greater the Participation in International Peacekeeping Missions, the Greater the Gender Inclusiveness

The second hypothesis in this category looks at the effect that a state's involvement in peacekeeping operations might have on the integration of women into the military. As discussed in Chapter 1, the United Nations has been trying to influence states to increase the percentages of women peacekeepers to better deal with humanitarian emergencies around the world, because 80 percent of refugees are women and children.[19] That women are increasingly participating in peacekeeping is clear from the latest data. According to the United Nations Department of Peacekeeping Operations, "between February 2007 and January 2008 there was an increase of over 40% of women serving in peacekeeping."[20]

But does the extent of the state's participation as well as the increased female participation in these operations have an effect on the IGI? First, I ran a correlation between IGI and the size of an individual state's personnel contributions to the United Nations Peacekeeping Operations. The mere volume of personnel contributed to the UN does not seem to reveal anything ($R = 0.302$, $p = 0.162$). There is no correlation between the two, which shows that just because the state contributes a large number of forces to various peacekeeping missions does not

---

18   Lindley-French and Algieri "Venusberg Report" (2009: 9).

19   Women's Commission for Refugee Women and Children (now the Women's Refugee Commission).   Available   at: http://www.womensrefugeecommission.org/about/714–20th-anniversary (accessed May 2, 2009).

20   United Nations Peacekeeping Factsheet February 2008.

mean it will be more inclusive of women. In addition, I ran a correlation between IGI and the percentage of women participating in military operations abroad in general. It also shows that just because the state sends proportionately more women soldiers to participate in military operations abroad does not mean it has integrated women to the highest degree (R = 0.023, p = 0.925).

The Longer the NATO Membership, the Greater the Gender Integration in the Military

The last set of hypotheses is about the relationship between IGI and the length of a state's NATO membership. The results demonstrate that there is no correlation between the two (R = 0.029, p = 0.893). Because the sample states all belong to the same military alliance, the best way to assess the impact that NATO Headquarters has on domestic policy changes is to test the correlation between the timing of the integration of women into the military and the timing of accession into NATO. The model shows a rather significant positive correlation between the two (R = 0.602, p = 0.002). Moreover, a closer look at the scatterplot shows clustering on two sides. This clustering tells us that the original member states of NATO (states ascended prior to 1989) on average integrated women much earlier than the new members (states ascended after 1989).

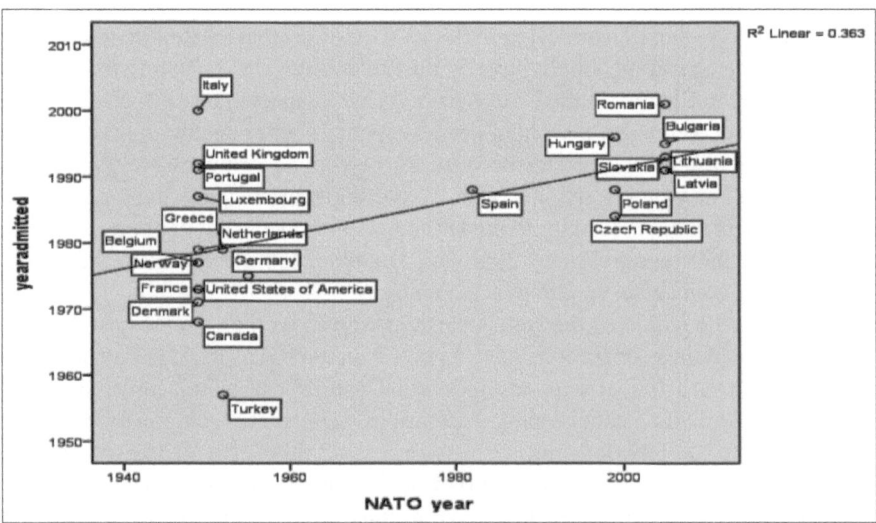

Figure 3.2 Year admitted and NATO year membership

In fact, the timing of NATO membership also explains 36 percent of the variation in timing of the integration of women into the military in individual member states.

Original versus New Membership (or East versus West)

The original "laundry list" has been reduced to five independent variables: conscription ratios, percentage of women in professional and technical fields, women's participation in the labor force, autonomy of women's movements, and timing of membership in NATO. The first three variables have not only shown a significant correlation to IGI in tests including all 24 states, but also explain an extremely high degree of variation in my dependent variable (R= 0.843, R squared = 0.710.). Together they explain 71 percent of the observed variability in the index of gender inclusiveness. This is a remarkable result. The "autonomy" variable has also shown to be a great predictor, but it was tested only in Western European and North American states. In those regions, it explains 43 percent of the variation in IGI (R = 0.653, R squared = 0.426)

The last variable, the timing of membership in NATO, is useful for another reason. It demonstrated the need to adjust the model to account for regional differences. Another reason to adjust the model is possible outliers. Because regression analysis assumes that variables have normal distributions, it is always necessary for visual inspection of data to check for outliers that can distort relationships and significance tests. That outlier has consistently been Turkey, the only member state with a majority Muslim population and the only member that is not considered a completely free and democratic state. In order to adjust the model, I have run it again without Turkey and I have controlled for the length of NATO membership.

In this model, I divided states into two categories, original Western European/ North American members and new Eastern European members. Then, I created a dummy variable where each group is coded as belonging to (1) and not belonging to (0).

In the original member states, the strongest and most significant correlation is between IGI and women in professional and technical fields (R = 0.676, p = 0.011), and alone it explains 48.7 percent of the variation. The next relationship that is also a strong, positive and significant correlation is between IGI and women's participation in the labor force (R = 0.622, p = 0.018). And the last strong determinant of IGI is the presence of an autonomous women's movement (R = 0.653, p= 0.011). This means that Western, highly developed and democratic states that have autonomous women's movements, a high number of women in professional and technical fields, as well as high percentages of women employed in general, have higher IGI scores. In addition, these three variables together explain 82.5 percent of variation in IGI values.

On the other hand, the new members from Eastern Europe seem to have very different reasons for integrating and expanding the role of women. The only, and relatively strong, correlation that the model reveals is between the Conflict Intensity Scale and IGI (R = −.710, p = 0.032). This negative correlation tells us that the lower the state's military operational capabilities and the lower their ability to fight full-scale warfare, the higher the gender inclusiveness. States such as Latvia, Lithuania and Hungary all have only general purpose ground forces and very high levels of gender inclusiveness, while others, such as Poland and the Czech Republic, have more sophisticated Special Forces and MEDEVAC capabilities, and lower levels of gender inclusiveness. This variable is a strong predictor and explains an overwhelming 50.4 percent of variation in gender inclusiveness levels in Eastern Europe (R squared = 0.504). These regional differences will be further explored in Chapters 4 and 5.

Conclusion

Analyzing the process of policy change in democratic states allows us to assess the validity of previous models and arguments that the degree of gender integration in the military is a reflection of cultural, socioeconomic and political equality of women and men in society as a whole.

By examining 24 NATO member states, I sought to determine which factors help us explain different degrees of gender integration. With the exception of Turkey, all of the states in my sample are considered fully democratic. I found that the primary causes of the change are rooted in the abolition of conscription, professionalization of the military services and technological advancements that no longer require most soldiers to be physically present on the battlefield. In addition, evidence shows that levels of gender inclusiveness in the military are not necessarily a reflection of gender equality in political life, measured as women's presence in politics and on ministerial levels, nor is it affected by cultural variables such as levels of religiosity or type of religion. However, this empirical evidence shows the danger of generalization, as it "unpacks" the complexity of factors affecting gender integration in the military in different regions. After repeating the test and separating the sample into two based on the timing of their NATO membership, the findings reveal different factors at work in each sample. States are more likely to admit women around the time of their accession into NATO; however, the length of that membership does not affect the degree of inclusion. While the existence of autonomous women's movements is useful in explaining the inclusion in Western Europe, different levels of military preparedness and operational capabilities are a better predictor in Eastern Europe. In order to get a better picture of the dynamics of policy process of gender integration into the military in old and new NATO member states, there is a need for a further, qualitative analysis.

Table 3.5 Summary of correlations

		All NATO	Original members	New members
Conscript ratios	Pearson correlation	−.587**	−0.425	−0.387
	Sig. (2–tailed)	0.008	0.130	0.303
	N	24	14	9
Unemployment	Pearson correlation	−0.282	0.056	0.000
	Sig. (2–tailed)	0.182	0.849	1
	N	24	14	9
Pop. growth 2009	Pearson correlation	−0.152	0.438	−0.365
	Sig. (2-tailed)	0.477	0.117	0.334
	N	24	14	9
Perc. pro. tech	Pearson correlation	0.482*	0.676**	0.483
	Sig. (2-tailed)	0.020	0.011	0.188
	N	23	13	9
Labor Participation	Pearson correlation	0.743**	0.622*	0.499
	Sig. (2-tailed)	0.000	0.018	0.171
	N	24	14	9
GNI	Pearson Correlation	0.284	0.419	0.241
	Sig. (2-tailed)	0.178	0.120	0.532
	N	24	15	9
Autonomy	Pearson correlation	0.689*	0.653*	
	Sig. (2-tailed)	0.005	0.011	
	N	15	14	
Perc. lower house	Pearson correlation	0.239	0.062	−0.144
	Sig. (2-tailed)	0.262	0.833	0.713
	N	24	14	9
Women ministers	Pearson correlation	0.240	0.254	0.276
	Sig. (2-tailed)	0.258	0.360	0.471
	N	24	14	9
Catholic	Pearson correlation	0.139	−0.009	0.000
	Sig. (2-tailed)	0.517	0.976	1
	N	24	14	9
Protestant	Pearson correlation	0.283	0.243	0.260
	Sig. (2-tailed)	0.180	0.402	0.500
	N	24	14	9
Islam	Pearson correlation	−.608**	a	a
	Sig. (2-tailed)	0.002	0	
	N	24		
Religiosity	Pearson correlation	−0.298	0.152	−0.445
	Sig. (2-tailed)	0.157	0.603	0.231
	N	24	14	9

		All NATO	Original members	New members
Men better executives	Pearson correlation	–.719**	–0.354	–0.468
	Sig. (2-tailed)	0.006	0.389	0.532
	N	13	8	4
Men better politicians	Pearson Correlation	0.276	0.105	0.153
	Sig. (2-tailed)	0.253	0.788	.694
	N	19	9	
Conflict intensity scale	Pearson correlation	–0.036	0.493	–.710**
	Sig. (2-tailed)	.879	0.159	.032
	N	21	11	9
UN troop rank	Pearson Correlation	0.302	0.144	0.668
	Sig. (2-tailed)	0.162	0.698	0.070
	N	23	14	8
Perc. fem. deployed	Pearson correlation	0.023	0.172	–0.073
	Sig. (2-tailed)	0.925	0.613	0.863
	N	19	11	8
Year admitted	Pearson correlation	0.147	–0.377	0.122
	Sig. (2-tailed)	0.494	0.184	0.755
	N	24	14	9
NATO year	Pearson correlation	0.029	0.131	0.274
	Sig. (2-tailed)	0.893	0.656	0.476

Notes: *Correlation is significant at the 0.05 level (2-tailed).
** Correlation is significant at the 0.01 level (2-tailed).
a. Cannot be computed because at least one of the variables is constant.

Chapter 4

Gender Integration in the Original Members of NATO: Case Studies of the United States and Italy

Introduction

In this chapter, I trace the process of gender integration within two original NATO members, the United States and Italy, to examine whether my argument captures the dynamics of policy change in individual cases. Today 15 percent of American armed forces are women, while females constitute only 3.9 percent of Italian forces. The U.S. government has undertaken a broad policy on gender integration since 1948, while the Italian government engaged in institutional discrimination by excluding women from the services until the year 2000. Therefore, the U.S. has among the highest levels of gender inclusiveness in the military (IGI = 16), while Italy has the lowest of all Western European and North American members (IGI = 8).

This chapter will illustrate that the quantitative analysis correctly identifies the major factors that affected gender integration in the military policies in United States and Italy. These were conscription, presence of independent women's movements, women's labor participation and women in professional and technical fields.

While the abolition of conscription affects the timing of the introduction of gender integration policies in both states, the extent to which integration took place in these two states is really determined by the other two factors: gender equality in the economic sector and the presence of autonomous women's movements. In the United States progressive women's groups were successful in framing gender integration in the military within a larger discussion of equal opportunity in labor force, forming strategic partnerships with congressional and service members sympathetic to the cause, and presenting a united front to the American public (Feinman 2000; Morden 2000). I contend that both the statistical analysis and the historical narrative of policy change in the United States support the argument that gradual gender integration in the military since World War II mirrors the increase in gender equality in the economic sector, as measured by the increase of women in the labor market and women in professional and technical fields in the civilian sector. The case of the United States not only confirms that there is a clear spillover of highly qualified, educated and skilled women from the civilian into military sectors but it also demonstrates that the greater integration in the military is largely dependent on the success of institutional reforms providing access and equal opportunity to women and men in the civilian labor market.

On the other hand in Italy, the legacy of the Fascist regime's cooptation of women's life and women's agenda, and the resistance movement's Communist call for class equality rather than gender equality are the main factors behind the lack of interest in the women's movement in a policy of integration into the military. Simply put, the Italian women's movement that declared "women were not to be defined in relation to men" interpreted this as yet another right-wing attempt at the militarization of Italian women's bodies (Malagreca 2006). In addition, Italian feminism in the 1970s developed largely through radical leftist political ideologies which the American public generally during that period found absolutely unacceptable. In the United States, radical left-wing parties were never crucial political players that could provide feminists with better access or be a significant political force behind feminist agenda.

On the other hand, there is no evidence to argue that the other proposed hypotheses are valid. First, among the military shortage variables, demographics and unemployment rates both fail to explain the difference in IGI scores between Italy and the United States. Italy has the lowest birth rates among all original members of the alliance, while the United States has among the highest. When it comes to unemployment, both states have relatively similar rates with the United States at 5.78 percent and Italy at 6.75 percent, yet very different IGI scores. This is exactly the opposite trend from the one that was proposed in Chapter 2.

Second, among domestic political and economic variables, there are three factors that have shown not to have an effect on the IGI in the quantitative analysis: the percentage of women in legislature, women at ministerial level, and the level of economic development. This seems to be true in both of my case studies. Italy has a higher percentage of women in legislature (21.3 percent) than the United States (16.8 percent), and both have exactly 24 percent of ministerial positions occupied by women. This is contrary to the original propositions, and demonstrates lack of the explanatory power of both factors. Although the Italian level of economic development is lower than that of the United States, it is still relatively on a par with the rest of Western European states that have much higher IGI scores.

Third, both cases help us reject the original argument regarding the effect of cultural variables on the level of gender inclusiveness in the military. Both Italy and the United States have much higher levels of religiosity than the majority of the original membership, as well as similar societal values regarding women in politics and business. Again, this confirms the findings from the quantitative analysis and refutes the original propositions.

And lastly, it was hypothesized that states with lower levels of threat and bigger contributions of forces to various peacekeeping missions around the world will have greater levels of gender integration. Although the U.S. armed forces have the highest level of military preparedness among all NATO members, Italian armed services are not far behind. Both states have continued to improve their services in the past decade. Again, this is contrary to what was argued by previous models. Similarly, the volume of peacekeeping contributions does not seem to be a useful predictor either. Italy is the highest contributor of all NATO states and has the lowest IGI score, while the United States ranks tenth in contributions, and has the highest IGI score.

Historical Background of Gender Integration in the US

Throughout much of American history, women have served in the military ranks as unofficial and temporary soldiers, nurses, spies, and support personnel. Having been ignored by many of the scholarly works on the Civil War, their stories were rescued from complete historical amnesia only in recent years by feminist historians seeking to set the record straight. Thanks to such works, today we know that women joined both the Union and Confederate ranks by assuming male identities (DePauw 1981; Leonard 1994, 1999; Blanton and Cook 2002; Tendrich-Frank 2007). While many were nurses, spies, and messengers, most of the women serving in the Civil War were members of the infantry engaged in combat alongside men, including five women who fought at Gettysburg.[1] It is estimated that about 400 women soldiers participated in the war.[2] However, as the conflict ended, women went home, and since they had officially not been part of the war, their efforts were largely forgotten.

During the Spanish-American War, the U.S. Congress allowed the military to recruit 1,500 nurses in the fight against typhoid fever in military camps, though they were considered civilian contractors.[3] As their presence on the field was crucial to the survival of the troops, the nation recognized the need for an ongoing presence of women military nurses. The creation of the Army Nurse Corps in 1901 and the Navy Nurse Corps in 1908 marked the official entry of women into the United States armed forces. As M.C. Devilbiss points out, "women served *with*, not *in* the armed forces" for much of American history, and the two Nurse Corps were still somewhat outsider and auxiliary organizations due to the fact that women did not have a rank or right to veteran's benefits. Finally, in 1917, women entered the Naval Coastal Defense Reserve as yeomen and the Marine Corps Reserve as clerks in 1918. These 12,500 Navy recruits and 305 Marines were the first women to hold actual military rank and status.[4] About 34,000 women served in World War I, including approximately 20,000 in the Army and Navy Nurse Corps.[5] After the war, the women of the Nurse Corps were kept active and the rest discharged.[6]

---

1 Blanton and Cook (2002: 15); De Pauw (1981).

2 Blanton and Cook (2002: 6); Holm (1982: 6).

3 Mann-Wall (2001: 56); for an in-depth historical account of Mercedes H. Graf, see Graf (2001).

4 Devilbiss (1990: 1–3).

5 Holm (1982: 10).

6 Devilbiss (1990: 4).

WAAC and WAC

The first attempt to bring women back was a bill proposed by Congresswomen Edith Nourse Rogers of Massachusetts in 1941 to create the Women's Army Auxiliary Corps (WAAC), which would allow women to attain the same military rank and privileges as men. The reason why Rogers was proposing this was because women who served during World War I received no benefits of official status, and had no legal protection or medical care. Once they returned home, they were not eligible for any veterans' benefits such as pensions or disability that were available to men. Rogers wanted to make sure that if American women were to serve again that they would obtain the same legal protection and be eligible for the same benefits as their male counterparts. However, the bill was regarded as too controversial by the War Department and was resisted for months, particularly by the southern congressmen. One of them asked: "Who will then do the cooking, the washing, the mending, the humble homey tasks to which every woman has devoted herself; who will nurture the children?" The bill was introduced by Rogers in May 1941, but it failed to be considered until after the Japanese attack on Pearl Harbor in December and until she gained support by the military leadership that understood the value of women in the military. Despite fierce opposition from generals such as George S. Patton, Jr. who was known to scream: "War is for men!" at his staff, others such as General George C. Marshall, the Army's Chief of Staff, understood that public sentiment toward women in uniform was changing and favored the creation of some form of a women's corps. Marshall argued that the United States would face a manpower shortage due to the global nature of the war, and fronts in both Europe and the Pacific. According to his analysis, the country could not afford to train men in clerical jobs, for which they were anyway inherently ill-suited due to their lack of patience. Hence, he was supportive of recruitment of highly skilled women who would provide such services more effectively and efficiently. Marshall worked closely with Rogers to write and sponsor the bill that would then represent a compromise between the two sides. Even though Rogers wanted the women's corps to be a part of the Army and to have equal benefits, Marshall and the Army were not ready to enlist women directly into its ranks. However, his support and congressional testimony helped push the bill through Congress. A week after the fall of Corregidor, and the capture and imprisonment of General Wainwright and 11,000 American soldiers, the Rogers bill was passed despite a bitter and prolonged debate that covers 98 columns in the *Congressional Record*. On May 14, 1942, Congress approved the creation of a Women's Army Auxiliary Corps (WAAC) "for the purpose of making available to the national defense the knowledge, skill, and special training of the women of the nation."[7] The final legislation was a compromise that created an auxiliary force in which women again not only lacked military status, but they also did not

---

7   Treadwell (1954: 19).

possess binding contracts and legal protections if stationed abroad. The Army was now responsible to provide up to 150,000 "auxiliaries" with food, uniforms, living quarters, pay, and medical care, but female officers would not be allowed to command men. Also, while women were not specifically prohibited from serving abroad, there was no overseas pay, government life insurance, veteran's medical coverage, and death benefits available to them. In addition, if captured, WAACs were provided no protections under existing international agreements regarding prisoners of war. Despite her intention to obtain pay, benefits and protection for women working with the military, Rogers was forced to make many compromises to get the necessary support in Congress. The following day, President Franklin D. Roosevelt signed the bill and set a recruitment goal of 25,000 for the first year. That goal was met within four months, and Secretary of War Henry L. Stimson immediately authorized WAAC enrollment at 150,000, which was in the original ceiling provided by the bill. Only a few months later, Eisenhower, who was initially not very supportive of women in the military, changed his mind while observing British service women during his preparation time for Operation Torch in London. After his "conversion," he had to talk to his generals and was acutely aware of the opposition he was facing as he argued that "these men had failed to note the changing requirements of war ... and an army of filling clerks, stenographers, office managers, telephone operators and chauffeurs had become essential, and it was scarcely less than criminal to recruit these from needed manpower when great numbers of highly qualified women were available."[8]

Feeling that she could gain more support, Rogers introduced another bill in 1943 to enlist and appoint women in the Army of the United States. Roosevelt signed the bill on July 1, 1943, and 90 days later the WAAC was discontinued and in its place came the Women's Army Corps (WAC), giving its members military status similar to that of the Women Accepted for Volunteer Emergency Service (WAVES).[9] Their utilization was seen as pure necessity and the best way to free up men for combat. The majority of the 350,000 women who served during World War II were employed in nursing, administrative and menial positions, using their skills from the civilian labor market. They were excluded from combat, and from all jobs requiring physical strength, all supervisory positions, and other roles such as personnel specialists and psychological assistants, as those, too, could be classified as "combat" positions."[10] By the end of May 1946, WAC strength had dropped from a wartime high of more than 99,000 to about 21,500 and by the end of May 1948, WAC strength totaled approximately 6,500 women on active duty.

---

8   Breuer (1997: 27).

9   Holm (1982: 24–27).

10   Holm (1982: 45); Devilbiss (1990: 7).

WAVES

Roughly two months later, what followed was the creation of the Navy Women's Reserve, in the form of the Women Accepted for Volunteer Emergency Service (WAVES), in July, 1942. Not many within the ranks were impressed with this change and one admiral was quoted as saying that he would have preferred "dogs or ducks or monkeys" (Breuer 1997: 21). The WAVES were not allowed on combat ships or aircraft, and could only serve in the continental United States until late in World War II when they were authorized to serve in certain U.S. possessions and Hawaii. The war ended before any WAVES could be sent to other locations. What made the WAVES different was that unlike the WAAC that served with the Army, the WAVES was considered an official part of the Navy, and therefore women held the same rank, pay, ratings and were subject to the same military discipline as male personnel. There were well over 8,000 female officers and almost 90,000 enlisted WAVES, serving as parachute riggers, radio dispatchers, clerks, mechanics, lab technicians, mail carriers, navigators, cryptologists, hospital corpsmen, linguists and weather specialists. By the end of the war, they accounted for 2.5 percent of the Navy's total strength. Most of the officers were restricted to the rank of lieutenant with the notable exception of Captain Mildred McAfee, the former president of Wellesley College and the WAVES director. Two other services relied upon the Navy's training facilities for both its officers and its enlisted women. The Coast Guard Women's Reserve (SPARs) was established on November 23 1942, and the Marine Corps Women's Reserve (USMCWR) on February 13, 1943. Two Navy restrictions were carried over to the Coast Guard. The SPARs could not serve at sea or outside the continental US, but, like WAVES, were authorized later to serve in Alaska and Hawaii. Although they were given the same pay, the SPARs had no authority over any man regardless of rank. About 11,000 women served as clerks, drivers, cooks, radio operators and operators of a confidential long-range electronic navigation system called LORAN. SPAR recruiting ended in December 1944 and was disbanded shortly after the surrender of Japan. Although a few SPARs were allowed to remain on active duty to finish the projects on which they were working, the remaining 12,000 were sent home and their records have been largely destroyed.

The USMCWRs received their commission based on their education and ability, and served as radio operators, drivers, aerial gunnery instructors, control tower operators, cryptographers and clerks. By the end of World War II, 85 percent of all enlisted U.S. Marine Corps personnel assigned to headquarters were women. The majority was demobilized at the end of the war, and of 20,000 only 1,000 remained by July 1946.

WASP

On September 1, 1942, in her column "My Day," Eleanor Roosevelt wrote:

> This is not a time when women should be patient. We are in a war and we need to fight it with all our ability and every weapon possible. Women pilots, in this particular case, are a weapon waiting to be used.[11]

According to William Breuer's account, it was Eleanor Roosevelt's words that prompted the Washington establishment to act fast, and only days later, on September 10, the Women's Auxiliary Ferrying Squadron (WAFS) was created to recruit female pilots whose job was to ferry aircraft from factories to U.S. Army Air Corps around the nation (1997). They had to purchase their own uniforms, that included grey-green slacks. Most Americans were unaware of these young women, who did not look "very military to civilians."[12] In one instance, Jack Dempsey's restaurant in New York City refused them service at first for wearing slacks, but ended up paying for their drinks and meals later after a male Air Force officer angrily reacted and explained that they indeed were military personnel. Similarly, after performing an emergency landing in Georgia, four female pilots were thrown in jail because a police officer mistook them for prostitutes. In addition, the Women's Flying Training Detachment (WFTD) was created only five days later, on September 15, 1942. What Roosevelt pointed out in her column was that women were already proving themselves a great asset to British war efforts, in which women pilots had been flying for the Air Transport Auxiliary (ATA) since 1940, and she saw no reason why the same could not be done in the United States. In fact, two American pilots, Jacqueline "Jackie" Cochran and Nancy Harkness Love, have already submitted and had their proposals turned down by General Henry Arnold prior to the Pearl Harbor attack. Cochran went to Britain to fly with the ATA, but returned to lead the WFTD, and Love became the WAFS commander.

The WAFS and WFTD were reorganized on July 5, and were now under the Women's Airforce Service Pilots (WASP). The WASPs were stationed at 120 air bases across the US, flew over 60 million miles during the war in every airplane. Thirty-eight were killed in the line of duty and 32 were injured. Unlike the WAVES and WAC, the WASP was considered a civilian service, and a fallen WASP was not a military hero, but rather a civilian whose body was sent home at family expense and without military honors. On September 30, Representative John Costello of California introduced the WASP Militarization Bill 3358, and on February 17, 1944, he introduced a longer version, now House Bill 4219, to give the WASP full military status and with it, insurance, burial benefits, and veteran status. Cochran spent years lobbying, and was supported by Arnold, who testified on Marcy 22, 1944 to the House Committee on Military Affairs that "we should use every means we can to put women in where they can replace men. The bill (House Bill 4219) will help to do that but will also make for more effective employment

---

11   Available at: http://libertyletters.com/resources/pearl-harbor/eleanor-roosevelt-women-pilots.php 9accessed February 13, 2014).

12   Breuer, 23.

of the present WASPs that we have in our service." That hearing lasted only one hour, and Arnold was the only person who testified. Arnold suggested both the militarization and expansion of the WASP, and showed evidence contrary to the argument that male civilian pilots and instructors who were affected by the Air Force training program cutbacks were being replaced by WASPs. In reality, these men just lost their draft-deferred status and were threatened with introduction into the Army's infantry. Defeating this bill was a way of avoiding the draft and combat service (Merryman 1998). Despite the fact that they did not meet physical and mental requirements for military pilots, like the WASPs, they led the opposition against the militarization bill. They were not alone as the press including the *Washington Post*, *Washington Star*, *Washington Daily News* and *TIME* magazine, treated the WASP as a silly, unnecessary and expensive program training inexperienced women. The bill was rejected on June 5, 1944 after the House Committee on the Civil Service, known as the Ramspeck Committee, concluded that the WASP was wasted money and wasted effort (Tanner 1996). On June 26, 1944, Arnold ordered for all training to end once the classes in session were completed. Six months later, on December 20 the WASP was disbanded, and all the records were classified and sealed until 1977 when President Carter signed the G.I. Bill Improvement Act, finally granting the WASPs full military status for their service.

Hence, the changing circumstances following the attack on Pearl Harbor and subsequent manpower shortage resulted in major changes in women's status. After the war, women's future role with the military was once again called into question. The central debate was between those who advocated permanent military status for women and those calling for a permanent women's reserve. Two powerful groups faced off over this issue. On one side were congressional leaders, claiming to speak for the American people, and on the other side were the commanders of the armed forces. At the center of this heated debate was Maine Representative Margaret Chase Smith, who pushed for legislation giving women regular military status. She was the first woman to serve in both the Senate and the House. The first meeting in which women's potential permanent military status was discussed took place within the House of Representatives Naval Affairs Committee in May, 1946, when debate resulted in no agreement, and a frustrated Chase Smith proclaimed the "Navy either needs these women or they do not."[13]

It was not until the Eightieth Congress (January 3, 1947 – January 3, 1949), when the Naval Affairs Committee and the Military Affairs Committee were merged to form the Armed Services Committee that the issue came up again. During Senate hearings, the military leadership, including Eisenhower and Admiral Nimitz, admitted that women were necessary because of their clerical skills and "nothing short of permanent status was an option."[14]

---

13   Sherman (2001: 69).
14   Sherman (2001: 70).

Shortly thereafter, the Women's Armed Services Integration Act was passed in the Senate in July 1947, but once it arrived at the Armed Services Committee in the House, the Act's progress came to a full stop. Committee Chairman Rep. Walter Andrews (R-NY) and a ranking minority member, Rep. Carl Vinson (D-GA), sat on the bill for more than six months before being confronted by another ASC member, Chase Smith.[15]

On February 18, 1948, the first hearing of the House Armed Services was convened. Among those present were Secretary of Defense James Forrestal, General Eisenhower, General Bradley, Admiral Denfeld, General Hoyt S. Vandenberg, Army Director of Personnel and Administration, General Paul, and his counterparts in the Navy, Marine Corps, and Air Force, Director and WAC Colonel Hallaren, and her counterparts in the other services.[16] Andrews openly expressed his antipathy towards women being in regular permanent forces just like men. On the other side of the debate representing integrationists, Eisenhower simply said: "I think it is a mistake to put [women] on a Reserve basis rather than a Regular. I think they should be an integrated regular part of the Army. I think the Air Forces feel the same way. We need them."[17]

The original bill was revised again giving women reserve status only and it passed with 26 yes votes, with a single no vote from Chase Smith. The bill was quickly listed by Andrews on the consent calendar, where most non-controversial bills are directed. After Chase Smith publicly complained and demanded unanimous agreement before the bill's listing, it was forced to the House floor. This way she would be able to publicly demonstrate and discuss the value of women in the armed services. She singlehandedly succeeded in the endeavor by making the members of committee confront the injustice they were attempting to commit against women by excluding them from permanent service. Without Chase Smith's forcing of this particular piece of legislation, women's integration into the military would have suffered a major setback.[18]

The delay in the passage of this bill also allowed other congresswomen, women veterans and major women's organizations that supported the legislation to organize and publicize their cause to the American people. Among these were Congresswomen Edith Nourse Rogers, WAVES Director and Captain Joy Bright Hancock, and WAC Colonel Hallaren, the Daughters of the American Revolution, the Women's Patriotic Conference on National Defense, General Federation of Women's Clubs, Women's Overseas League and National Civilian Advisory Committee for the Women's Army Corps.[19]

Major newspapers came out fighting the House ACS and sided with the military leadership. The *New York Times* called for women's integration with full

---

15 Sherman (2001: 70–71).
16 U.S. Congress (1948: 5565).
17 U.S. Congress (1948: 5566).
18 Morden (2000: 51–52); Sherman (2001: 71–72).
19 Holm (1982: 115); Morden (2000: 53).

military status and declared that "to dispense of their services would be a sorry blunder. Replacements for them simply cannot be found, with the shortages that now exist in all ranks."[20] The *Washington Post* called the failure of the House ASC "a shortsighted waste"[21] and the *Christian Science Monitor* concurred by warning of "weakening of national defense by assignment of able-bodied men to jobs that women usually perform more efficiently."[22]

Finally, after years of debate and numerous changes made to accommodate both armed services and congressional members, on June 2, 1948, Public Law 625, the Women's Armed Services Integration Act, was passed. It was signed into law by President Truman 10 days later. The WAC was now a permanent and separate organization within the Army. Although the Act guaranteed women permanent status in the military services, it placed highly specific restrictions on women. They could make up no more than 2 percent of the total enlisted ranks. The proportion of female officers, excluding nurses, could equal no more than 10 percent of enlisted women. No woman could serve in a command position, reach the rank of general, or hold permanent rank above lieutenant colonel or Navy commander. It established separate promotion lists for women officers in the Army, Navy and Marine Corps. Only the Air Force had an integrated promotion list for ranks below colonel. This Act specifically prohibited women from being assigned to aircraft or ships engaged in combat missions. Because the Navy and the Air Force have the most ships and aircraft, this Act applied most directly to them. However, the Secretary of the Army developed policies to exclude women from direct combat, based upon the implied congressional intent of the Navy and Air Force statutes.[23]

Women's Groups Inside and Outside Government

USA

During the 1950s and early 1960s, nothing changed regarding gender integration in the military. However, this is not to suggest that nobody was working on the change. Shortly after women were integrated in 1948, Secretary of Defense Marshall, at the urging of his Assistant Secretary of Defense for Manpower, Anne Rosenberg, helped create the Defense Advisory Committee on Women in the Services (DACOWITS).[24] It was to be composed of civilian women "to provide

---

20   "Women in the Services." *New York Times* (1857–Current file), March 31, 1948; ProQuest Historical Newspapers, the *New York Times* (1851–2005), 24.

21   "Women and Defense." *The Washington Post* (1877–1954); January 7, 1947; ProQuest Historical Newspapers, the *Washington Post* (1877–1992), 8.

22   "Why Bar Women?" *Christian Science Monitor* (1908–Current file); March 29, 1948; ProQuest Historical Newspapers, *Christian Science Monitor*, 16.

23   Collier (1991: 3); Holm (1982: 119–120).

24   Mitchell (1998: 8).

advice and recommendations on matters and policies relating to the recruitment and retention, treatment, employment, integration, and well-being of highly qualified professional women in the Armed Forces."[25]

DACOWITS's first recruitment campaign was not a success, managing to recruit only 6,000 women, far less than the anticipated 72,000.[26] Holm called it a "disaster ... ill-conceived and very poorly timed."[27] DACOWITS's situation was exacerbated by the Korean War, whose horrific images only turned women away from joining. After the war, the numbers started to rise again, and this time DACOWITS stopped focusing on recruitment and instead directed its energies toward the general improvement of women's life in the military and the elimination of rank restrictions. They argued that women's components "reached maturity which calls for re-examination of the structure with respect to the maximum career potential afforded to new recruits."[28] And this time DACOWITS's efforts were well-timed as they coincided with a more widespread political protest against discrimination based on sex. The passage of the 1964 Civil Rights Act gave women another opportunity to fight to eliminate the legal barriers imposed by the Women's Armed Services Integration Act of 1948. The following year, President Johnson eliminated discrimination within the executive branch. Although this executive order did not directly address the military services, the Department of Defense decided to finally heed DACOWITS's advice and submitted legislation to remove percentage caps.

It is here where the real accomplishments of DACOWITS lie. They fought the battle on many fronts and succeeded. Holm describes their efforts and how:

> committee members pulled out all the stops—soliciting support from women's groups, encouraging letter-writing campaigns, focusing media interest, and individually lobbying Congress ... [they] held regular strategy planning sessions with military women; after each DACOWITS meeting the members fanned out over Capitol Hill, paying court to whomever they knew, gaining support for the legislation. Many had political connections in the White House and on the Hill, others direct access to the media, which they used.[29]

Twenty years after the initial Women's Armed Services Integration Act and seven years of the DACOWITS campaign, President Johnson signed Public Law 90–130 on November 8, 1967, removing the 2 percent cap on enlisted women and lifting

---

25   DACOWITS official website factsheet. Available at: http://www.defenselink.mil/dacowits/tableabout_subpage.html (accessed May 23, 2009).

26   Mitchell (1998: 9).

27   Holm (1982: 152).

28   Mitchell (1998: 12).

29   Holm (1982: 199); Also quoted by Mitchell (1998: 13–14).

promotion restrictions. The president declared: "There is no reason why we should not some day have a female Chief of Staff or even a female Commander-in-Chief."[30]

The enthusiasm did not last long and membership withered away with the Vietnam War and the radicalization of the civilian membership of DACOWITS that some accused of not knowing enough about the military. This was followed by the forced opening of DACOWITS meetings to the public, where other feminist organizations joined the discussion, among them the National Organization of Women and the Women's Equity Action League.[31] Authors like Mitchell declare that from here on "the committee's recommendations to the Defense Department were merely modulated renditions of the demands of professional feminists."[32]

While he may not be wrong in his assessment, Mitchell portrays these feminists as wicked witches who ruined the party for the boys. No matter how nasty and biased, this argument, in fact, shows that the autonomous and liberal women's movement did have an impact on gender integration in the military, and a positive one at that. Working with organizations outside the Pentagon meant stronger demands and better results for both sides—the U.S. Armed Forces and women. It shows that the women of DACOWITS and the "second wave" outsider feminists organized, disseminated information, and lobbied in order to amplify those voices seeking a greater integration of women into the military. While DACOWITS members were only concerned with the military, for the outsiders this was just another battle for equal opportunity in the workplace and equality in citizenship between men and women.

The American "second wave" feminists framed women as a minority group and this perspective focused on equal opportunities and emancipation as opposed to liberation and class struggle of the radical left wing. American feminists embraced classical liberalism that focuses on the rights of the individual, equal rights of all citizens, and property and privacy issues largely based on Locke's ideas of equal rights and protection.[33]

In 1971, the National Organization for Women started actively debating the role of women in the military and issued a resolution requesting the cessation of sexist practices in the military and the sexist basis for compulsory service.[34] Although DACOWITS did not press the issue of combat participation right away, by 1974 it was demanding a repeal of combat exclusionary laws and the opening of military academies.[35] In 1975, Public Law 94–106 admitted women into federally funded military academies and a year later the first women entered these institutions. In

---

30   Holm (1982: 192).

31   Moore (1996: 124).

32   Mitchell (1998: 113).

33   Whelan (1995: 27).

34   National Organization for Women and War Resolution adopted at the National Conference in 1971. Available at: http://www.now.org/issues/military/policies/war.html (accessed May 24, 2009).

35   Mitchell (1998: 113).

addition, now that the draft had ended with the expiration of the Selective Service Act and the All-Volunteer Force era had started, recruiting goals for women began to increase.

President Jimmy Carter's administration expanded the membership of DACOWITS, and within a week of being part of the government, Secretary of Defense Harold Brown requested a report on women in the military. The *Use of Women in the Military* was completed in May, 1977, and was followed up by the publication of *America's Volunteer* in December, 1978, both with the goal of furthering women's role in national security by increasing their numbers to 80,000, or 11.4 percent of the total force. In 1978, the WAC was abolished as a separate institution within the Army, and women in the Navy were assigned to non-combatant ships and temporarily on any ship for a period of less than six months.[36]

But as President Ronald Reagan's administration came to power, it was clear that progress was to not only decelerate but that it actually went into reverse. Holm argues that it was not a secret that many in the ranks and in the government "believed that military policy decisions were being made by well-meaning amateurs ... motivated more by political expediency and misguided desires for social equity than by the requirements of national defense."[37] In fact, the Army "decided that the time was right to roll back the Carter program"[38] and started to plan its reductions only a month after Reagan took office, by preparing to level out women's numbers to 65,000 instead of following Carter administration's recommendation of 80,000. In addition, the Army hinted at a new draft to fill 100,000 positions necessary for Reagan's new military build-up. Instead, the services were prevented from proceeding by Secretary of Defense Caspar Weinberger, who told them to eliminate institutional barriers.[39] Only a year later, Weinberger declared that "the most rewarding development we have seen in our armed forces over the past decade has been the tremendous expansion of opportunities for women."[40]

The Reagan administration added new members to DACOWITS. The organization now had some Republican women, who, regardless of their ideological stands, continued to pursue the feminist agenda within the military. For these new members, "being strong Reagan supporters was not synonymous with a conservative perspective on gender discrimination."[41] In May 1981, the Army commissioned a study to review the assignments of women to physically demanding positions. The report presenting the results of this study, known as Women in the Army Policy Review (WITA), made recommendations regarding Army personnel policies as they related to mission, combat readiness, quality of life aspects, and the utilization of female enlisted soldiers in the Army. The main finding was that

---

36   Rostker (2006: 563–564).
37   Holm (1982: 387–388).
38   Rostker (2006: 565).
39   Rostker (2006: 567–568).
40   Rostker (2006: 560).
41   Katzenstein (1998:74).

only 3 percent of women are able to perform heavy lifting but they occupied 42 percent of positions requiring the skill.[42] Both the Women's Equity Action League (WEAL) and DACOWITS fiercely opposed the findings, and demanded review of the study. They argued that not only was the methodology wrong,[43] but that such a study was negatively affecting the moral of women in the ranks, and was going to deprive the Army of the available and skillful recruits.[44] In October of 1983, the Army decided to reconsider the study to correct methodological errors. DACOWITS and WEAL members succeeded in pushing for the reopening of 13 out of 23 positions that WITA closed to women.[45] Women were to be employed in all positions except those explicitly prohibited by combat exclusion statutes and related policy.[46]

As a result, in November 1983, the Secretary of the Army implemented the Direct Combat Probability Coding (DCPC) system as a new Army policy specifying where women may be assigned on the battlefield. The DCPC classifies each position in the Army a ranking from P1 to P7 based on the probability of routine engagement in direct combat and women only do not have access to those positions that are coded P1. This system is informally known as the "combat exclusion" rule. However, only a couple of years later, in 1985, a new study was commissioned by the Commander, United States Army Training and Doctrine Command (TRADOC) to review the impact of DCPC policy guidelines, because there seemed to be a wide variety of interpretation and application of DCPC guidelines due to concerns which were not directly addressed by the policy. Hence, the Women in Combat Task Force (WCTF) study group was established and it was conducted by the Combined Arms Integration Directorate of the Combined Arms Combat Developments Activity, in Fort Leavenworth, Kansas. It was undertaken to evaluate the Army policy governing assignment and utilization of women soldiers in the combat zone, focusing on the DCPC policy with a primary purpose "to determine if changes are required to current Army policy governing utilization of women in combat."[47]

Data and information were included from research of published and 18 unpublished studies and reports. The WCTF study group also led field briefings with five division commanders and two major Army command commanders and their staff regarding the assessment of study's initial findings as well as performance of female soldiers within those commands. The study group conducted surveys with 1,102 personnel at the Command and General Staff College, Combined Arms Service Staff School, and the Sergeants Major Academy and held interviews

---

42   Levin (1987: 231).

43   Part of their argument was that the sample of women soldiers at Fort Jackson was too small to make such broad generalizations regarding physical abilities.

44   Mitchell (1998: 120).

45   Mitchell (1998: 122).

46   Rostker (2006: 569).

47   U.S. Army (1986: v).

with selected groups of Army personnel. In addition, the WCTF study group held workshops with TRADOC Proponency offices and branches and visited U.S. Army Forces Command (FORSCOM) and TRADOC posts and units. The three criteria developed to measure the validity of the WCTF's study group were: findings must support combat effectiveness, be consistent throughout the Army in the assignment policies of women, and be consistent with both, tenets of AirLand Battle and with the threat assessment.[48] The conclusion of the report was that the DCPC guidelines were sound, and required no changes. This study was groundbreaking, even though it did not openly support opening of the combat positions to women, but in a sense that it was evidence-based scientific inquiry that demonstrated that women's presence has no effect on combat unit cohesion and operational effectiveness: "Except for this combat exclusion policy, there are no sound reasons—practical or cultural—to categorically deny women assignments anywhere on the AirLand Battlefield as long as they are qualified to perform the required duties."[49] Although somewhat indirectly, it also emphasized the need to train women for combat if supporting combat units by arguing that: "Women cannot be protected from the enemy on the AirLand Battlefield and, in fact, must participate equally with men in killing or destroying enemy forces as necessary to defend themselves and their unit."[50]

As such, new reports were encouraging action, the work of women's organizations continued throughout the 1980s and 1990s, and new initiatives started to spring up outside the walls of the Pentagon. On the evening of December 7, 1982, at an event sponsored by WEAL, Dr. Linda Grant De Pauw thought of publishing a journal on women and war and women in the military. Shortly afterwards, she and a small group of other women established the MINERVA Center, which today is the most important sources of information on women in the military. MINERVA originally published the *MINERVA: Quarterly Report on Women and the Military*, a scholarly journal, and *Minerva's Bulletin Board*, a news magazine focusing mostly on American women in service and women veterans, which also carried a number of stories and news relating to women and the military around the world. The news magazine was replaced in 1995 by *H-MINERVA*, a listserv associated with the H-NET consortium of scholarly discussion groups.[51]

In 1987, DACOWITS visited Navy and Marine posts in the Far East and came back with a list of problems ranging from lack of promotion opportunities to sexual harassment. The report described the ships where abuse and disrespect were encouraged, incidents of which the Navy was clueless. Written by DACOWITS

---

48   U.S. Army (1986: viii).
49   U.S. Army (1986: x).
50   U.S. Army (1986: xi).
51   As a member of H-MINERVA listserv, I have enjoyed a privilege of learning about the history of the Center, as well as collaborating with the members whose work on women in the military spans more than three decades. Their resources were invaluable to this book.

Chairwoman Jacquelyn Davis, the report was not only submitted to congressional hearings but was circulated in the media by the Women and Military Project[52] of WEAL. In fact, this organization quickly became the main source of information for reporters, congressional offices, DACOWITS and military women. In terms of lobbying, WEAL was the main organization supporting gender equality in the military ranks.[53]

The task force created by Weinberger to investigate DACOWITS's complaints also found "problems concerning the implementation of the combat exclusion rules ... [and that] sexual harassment was a significant problem in all Services."[54] It found that combat exclusion rules were very problematic and difficult to interpret and as a solution, in 1988, the Department of Defense Task Force on Women in the Military recommended the "risk rule," which barred women from areas of the battlefield where the "risk of exposure to direct combat, hostile fire, or capture is equal to or greater than that experienced by associated combat units in the same theater of operations."[55] Although by now the government did open 63 percent of all positions in the military to women, it still severely limited their possible assignments, and the National Organization of Women mobilized to actively fight for a place for women in combat. In 1990, the NOW issued a resolution to fight this "inequality of opportunity." It demanded "equality for women in joining the military and in training, job assignments and benefits in the military; and ... that NOW actively supports elimination of statutory restrictions on women in the military."[56]

Progress continued in 1989 when two women commanded units in Panama. In 1990, the first female commanded a Navy ship; and in 1991, in the Persian Gulf War, large numbers of women moved forward with their units into combat zones, which they were officially forbidden to enter. This progress prompted more questions regarding the meaning of combat exclusion. Representative Patricia Schroeder, a Colorado Democrat and an advocate of rescinding all rules excluding women from combat, argued that "the Persian Gulf helped collapse the whole chivalrous notion that women could be kept out of danger in a war. We saw that the theater of operations had no strict combat zone, that Scud missiles were not gender-specific—they could hit both sexes and, unfortunately, did."[57]

---

52   The project was transferred later to the Women's Research and Education Institute and it continued being a crucial source of information for the congressional members. This group has done a great job facilitating exchange of ideas and expertise between feminist groups, equal opportunity networks and policymakers.

53   Kazenstein (1998: 67).

54   Rostker (2006: 570).

55   Rostker (2006: 571).

56   This resolution was adopted by the organization's National Board in September 1990. The text of the entire resolution is available at: http://www.now.org/issues/military/policies/wim.html (accessed May 24, 2009).

57   Nordheimer (1991).

Hence, in the aftermath of the Gulf War, the biggest issue on the Pentagon's agenda became women's role in combat. While General Vuono of the Army and General Gray Jr. of the Marine Corps were completely opposed to women soldiers in combat, Admiral Kelso Jr. of the Navy and General McPeak of the Air Force opposed the issue personally but were willing to support it if legislative changes were made. But women's achievements, including the fact that more than 40,000 women participated in the Gulf War, in which 15 were killed, pushed the Senate Armed Services Subcommittee on Manpower and Personnel to consider repealing the ban on women in combat in June 1991. Defense Secretary Dick Cheney supported the House legislation, which had passed only a month earlier, and allowed services to have women in combat, though they were not required to do so. Cheney's Assistant Secretary of Defense for Force Management and Personnel, Christopher Jehn, agreed but he also requested "maximum flexibility in regulating women in combat" so that it did not affect combat readiness and effectiveness of the troops as "it has a lot to do with the behavior of the men … and whether we are psychologically ready for it or not." Although many senators agreed, Sen. John McCain (R-Ariz.) worried that this "blanket denial of combat roles for women strikes many people as outdated and unfair."[58]

The years 1992 through 1994 saw considerable legislative and policy changes regarding the roles of women in the military as DACOWITS and congressional members such as Patricia Schroeder pushed to end the restrictions. The WEAL Women and Military Project was now part of the Women's Research and Education Institute under the direction of Carolyn Becraft, who successfully organized the lobbying strategy. Feminists inside and outside the military worked together to make it happen. While the uniformed women, particularly women pilots, could not directly lobby congressional offices, they went along with civilians to provide information and firsthand accounts in response to any question that those on the Hill might have.[59]

Shortly thereafter, in 1991 Congress allowed women on combat aircraft assignments and established a Presidential Commission on the Assignment of Women in the Armed Forces that spent the following seven months studying the possibility of opening more positions to women.[60] The commission's report was criticized by women's advocates, such as the National Women's Law Center, as "strongly biased against increasing assignment." Critics pointed to a number of weaknesses. Complaints included the use of anecdotal rather than empirical evidence, concern with problems outside the military effectiveness scope, little study of negative implications of current limitations, as well as problematic membership of the commission that included some well-known opponents of expanding women's role in the military.[61]

---

58 Schmitt (1991).
59 Katzenstein (1998: 50–51).
60 Godson (2001: 281).
61 Sagawa and Campbell (1992: 7).

The Presidential Commission recommended allowing women to serve on all surface combat ships except amphibious vessels but keeping them off the crews of combat aircraft and out of ground combat units. However, as a new administration arrived, the issue of women in combat was quickly taken up by both President Clinton and his Secretary of Defense, Les Aspin. After the "Risk Rule" was rescinded on January 13, 1994, at a special news briefing at the Pentagon, Aspin said that "expanding the roles for women in the military is the right thing to do, and it's also the smart thing to do. It allows us to assign the most qualified individual to each military job, which is very, very important when what we really rely on is the high quality of our personnel." In April 1993, Aspin sent a memorandum to the services asking them to give women the opportunity to compete for assignment to combat aircraft and to open as many Navy ships to women as possible under the legal restrictions that still prevented them from being assigned to ships engaged in combat missions. Since World War II, women had served on a variety of combat support vessels but were prohibited from assignment to destroyers, aircraft carriers and other fighting vessels. Moreover, Aspin's memorandum asked the Army and the Marine Corps to "justify the continuing exclusion of women from ground infantry units" but did not prescribe additional gender integration in these two services. Instead, it reaffirmed the exclusion of women from units below brigade level whose primary mission entailed direct combat on the ground.[62]

Additional changes in law and policy regarding the assignment of women were finally implemented in 1994, opening between 15,000 and 20,000 jobs previously off limits in the Army and Marine Corps. Troops in ground combat still excluded women, but Aspin replaced the "risk rule" with a new definition barring women from direct ground combat. According to this definition, women could not be in jobs engaging an enemy in close combat while being exposed to hostile fire and to a high probability of direct physical contact with a hostile force's personnel.[63] Basically, this memorandum directed that women be assigned to all units except those "below the brigade level whose primary mission is to engage in direct combat on the ground."[64]

Debate regarding the extent of gender integration continues, particularly as wars in Iraq and Afghanistan have exposed women to combat more than any other conflict, but the limitations persist. In May, 2005, the now-disgraced Republican representative from San Diego, Duncan Hunter, proposed limiting the role of women in combat once again. In the end, House Republicans abandoned their own plans after the original bill ran into opposition from the Pentagon and lawmakers from both parties.[65] For others, such as retired Air Force Lieutenant Colonel Karen

---

62 Lancaster (1993).

63 Schmitt (1994: A22).

64 Rostker (2006: 576).

65 "House Rejects Limits on Women in Combat." Associated Press, May 25, 2005. Available at: http://www.foxnews.com/story/0,2933,157607,00.html (accessed May 30, 2009).

Johnson, now executive vice president of the National Organization for Women, serving in combat positions is the duty of every citizen, including women. She argues that "to be an American citizen, you have a right to be able to participate in the military and to serve the military as a full and responsible citizen. As women, we are patriotic. We care about our country. And we're willing to fight and to die for our country."[66] And they do die for their country. Up to 2008, 97 women soldiers and 3 female civilian contractors were killed, including 61 by hostile action.[67] An additional 14 women were killed before the war ended.[68] The number of female soldier deaths represents a significant increase from previous wars. Two women were killed in the Korean War, 8 in Vietnam and 15 in the Persian Gulf War.[69]

From 2002 until 2008, 191,500 women have served, which is roughly 11 percent of all troops in Iraq and Afghanistan. In a 2007 report, the RAND Corporation demonstrated that although the Army was complying with the DOD assignment policy, hundreds of female Army members had received a Combat Action Badge, suggesting the Army recognizes that women have indeed been in combat.[70] With their numbers increasing and their role expanding, problems they face when they get home are also on the rise. These women have been more likely to suffer mental problems, post-traumatic stress disorder (PTSD) and homelessness, and many times services are not adequate. There are not enough clinics to deal with brain injuries regardless of sex, and there are not enough beds in mental hospitals to accommodate all who require help. Although women do not serve in combat, they still suffer physical and psychological injury, and today PTSD, hypertension, and depression are the three categories diagnosed most frequently among female veterans. But it is necessary to remember that male soldiers have experienced similar problems in recent years. Nearly 35 percent of all soldiers needed therapy a year after the war in Iraq started and today nearly one in five soldiers suffer mental problems.[71] But there is a concern in the ranks that is mostly a women's problem, and that is sexual harassment, which continues to plague the armed forces.[72] One in five women seen by Veterans Administration (VA) hospitals respond "yes"

---

66   Brian Daks (2005).
67   "100 Female U.S. Service Members Have Died in Iraq." CNN, July 24, 2008. Available at: http://edition.cnn.com/2008/WORLD/meast/07/24/iraq.main/index.html (accessed May 30, 2009).
68   Norland (2011).
69   Fisher (2005).
70   Harrell et al. (2007: xvii).
71   United States Army Medical Department Press Release Mental Health Advisory Team V Report. Available at: http://www.armymedicine.army.mil/news/releases/20080306mhatv. cfm?m=3&y=2008 (accessed May 31, 2009).
72   I say "mostly" because in 1995, the U.S. Department of Defense survey showed that 35.8 percent of males have also experienced sexual harassment behavior in that year alone. For more details see Heather Antecol and Deborah Cobb Clark (2001: 3–18). It is important to note the report also says that "Though rates of MST are higher among women, there are almost as many men seen in VA that have experienced MST as there are women. This is because there are many more men in the military than there are women."

when screened for Military Sexual Trauma (assault or harassment experienced while in the military).[73] That means 20 percent of Iraq and Afghanistan female veterans have also been victims of sexual assault and/or sexual harassment.

The effort by the women's groups both inside and outside the military institution focused solely on gender discrimination as opposed to a whole array of issues ranging from racism to homosexuality allowed this coalition to succeed.[74] Eliminating sexual harassment has from the beginning been under the umbrella of gender discrimination and these women's groups continue to fight it. It is important to understand the gravity and impact of sexual assault in the military in order to understand problems of recruitment, retention, and ultimately the effectiveness of female troops who are under constant pressures, as these are used as measurements for the future expansion of their role. The most recent study by Helen Benedict shows that a third of U.S. female soldiers have been sexually assaulted or raped, while 71 percent to 90 percent have been sexually harassed.[75] In 2010, fewer than 21 percent of cases went to trial, and only half resulted in conviction. The Department of Defense Sexual Assault Prevention and Response Office budget was roughly $23 million in 2010, which is more than four times what it was in 2005. But it is important to emphasize that the number of violent sexual assaults has also dramatically increased, and some estimate by a 64 percent since 2006. According to the most recent report by the Pentagon's Sexual Assault Prevention and Response Office, 3,192 sexual assaults were reported between October 1, 2010 and September 31, 2011—roughly nine per day. Such data are not as horrifying as the fact that the Department of Defense estimates that only 14 percent of assaults are actually being reported. That means the real number is approximately 19,000 per year or 52 assaults per day, with majority of women being attacked by their fellow soldiers. In fact, even former Defense Secretary Leon E. Panetta acknowledged that the statistics are much higher because many cases are covered up. The Defense Department's report shows that one in three women in the military has been sexually assaulted, compared with one in six in civilian women. In addition, most of the victims are female junior enlists under the age of 25, while their attackers are males, under the age of 35 and higher-ranking. And very few have reported the attacks. Some fear retribution, and others claim that "it is a career ender to come forward," so many women keep quiet in order to advance.[76] Despite the fact that Panetta called it an "outrage," rates of prosecution and conviction remain low, and there is no decrease in sexual assault; 489 subjects

---

73   United States Department of Veterans Affairs National Center for PTSD, Military Sexual Trauma. Available at: http://www.ptsd.va.gov/public/pages/military-sexual-trauma-general.asp (accessed November 10, 2012).

74   Katzenstein (1998: 75).

75   Benedict (2008).

76   Risen (2012).

had court martial charges initiated against them but of 240 subjects whose cases proceeded to trial, 80 percent were convicted. That means that only 6 percent of the total reports resulted in a conviction by court martial.[77]

Some argue that lowered recruitment standards and criminal waivers allowed more than 100,000 people with criminal pasts to join the military. According to the Palm Center, a study, "Gang-Related Activity in the U.S. Armed Forces Increasing," released in 2007 by the Federal Bureau of Investigation shows that the number of gang-related incidents in 2006 was 10,309.[78] The numbers coincide with the Army admitting about 20 percent of recruits in 2006 with a record of legal problems ranging from felony convictions and serious misdemeanors to drug crimes and traffic offenses. In 2007, three out of 10 recruits had a waiver. It was only in 2009 that the military met its recruitment targets for the first time since 2004, and in 2011 there were no recruits with misconduct convictions or drug or alcohol issues.[79]

In 2008, Rep. Louise Slaughter (D-NY) and Rep. Jane Harman (D-CA) provided testimony to the House Oversight and Government Reform Subcommittee on National Security and Foreign Affairs outlining the abuses and lack of punishment for the abusers. Slaughter pledged to introduce a bill that would reduce violence against military personnel and their families, establish a victims advocate office within the Department of Defense that would be responsible for providing services to victims, protect those who report the assaults, and court martial those who violate "no contact" or protection orders. Moreover, all branches of the services would have to provide victims advocate programs and train recruits in violence prevention. The Department of Defense asserted that SAPRO, the Sexual Assault Prevention and Response Office, an organization responsible for sexual assault policy, already exists. Michael Dominguez, principal Under Secretary of Defense for Personnel and Readiness, ordered the SAPRO director Dr. Kaye Whitley to ignore the congressional subpoena to testify.[80]

After congressional members expressed their outrage, Whitley was forced to testify in January 2009, when she admitted that "even though she was in a car in front of the Rayburn House Office Building, testimony in hand, ready to walk into the hearing ... understood the implication of ignoring a subpoena but was accustomed to following orders"[81] and therefore failed to show up. The following month Slaughter reintroduced the Military Domestic Violence and Sexual Assault Response Act (H.R.840) in the House. The National Organization for Women has taken action and is actively working on lobbying congressional members. As of

---

77   Department of Defense. *Annual Report on Sexual Assault in the Military, FY 2010–2011*. April 2011, 43–44. Available at: http://www.sapr.mil/index.php/annual-reports (accessed October 15, 2012).

78   The Palm Center (2008).

79   Baldor (2012).

80   Wadding (2008).

81   Cavallaro (2008).

April 2009, they had gathered 30, or one-third, of votes necessary to push the bill through the committee hearing and to the House floor, but it died and was referred to Committee. Meanwhile, the women of the United States military services are suffering in numbers higher than ever.[82] Almost two years later, Iowa Rep. Bruce Braley (D-IA) introduced a similar bill, the Holley Lynn James Act, to develop a Department of Defense sexual assault prevention and response policy. Holley Lynn James was a Second Lieutenant in the Army who was murdered by her husband, John Wimunc, a Marine, in July 2008. Nearly three years later, the chances of it passing appear very slim.

As for DACOWITS, the most important organization supporting women in the military, the story was almost over. President George W. Bush allowed their charter to expire in 2002, and by 2008 it had been allowed to almost disappear. DACOWITS had only five members, 35 less than it boasted in 1951. In December 2008, the National Women's Law Center (NWLC) and Women's Research and Education Institute (WREI) issued recommendations to President Obama's Transitional Team asking him to re-establish DACOWITS under the Federal Advisory Committee Act with a mission to advise the Secretary of Defense on a full range of matters and policies relating to women in the armed forces; to again serve as a vital link between the civilian community and the Department of Defense; to increase its membership to a minimum of 25 to be chosen from diverse backgrounds and geographic areas; and to allow the committee to focus again on military women's issues.[83] Today, there are 17 members, a far cry from what is needed to address the issues women in the ranks face.

As the influence of DACOWITS waned, so did the progress of women's integration during the eight years of the Bush administration. Although there are other military women's and veterans' associations and groups around the country for military women to exchange information unique to their experience and to offer advice and suggestions to those women who are considering joining the United States armed forces, nobody has managed to lobby or stand for U.S. military women the way DACOWITS did. Instead, during the Bush administration, groups opposing women in combat, and even joint training, have gained strength by joining forces with the right wing and the Christian right. Among the most outspoken critics of mixed-gender basic training, gays in the military, and women in combat is Elaine Donnelly, president of the Michigan-based Center for Military Readiness. She and 15 other advocacy groups have been seeking to end equal opportunity in training, and prevent it from ever taking place in combat because

---

82   Gandy (2009).

83   Recommendations from joint letter to Presidential Transition Team jointly signed by Women's Research and Education Institute & National Women's Law Center "DACOWITS Should Be Revitalized." December 8, 2008. Available at: http://www.nwlc. org/pdf/DACOWITS.pdf (accessed May 23rd, 2009).

it is "being imposed on the military by liberal activists"[84] In 2001, Donnelly made the following statement that tells us just how limited her appreciation of combat is:

> The DACOWITS and Pentagon feminists keep demanding that all-male, tip-of-the-spear combat units—the ones that directly and successfully engaged Taliban fighters on the ground-be made politically correct by including women. Never mind that the job of Special Operations helicopter pilot, as vividly portrayed by the real-life character Michael Durant in the film Black Hawk Down, is not just another "career opportunity."[85]

A year later, Donnelly was awarded a Conservative Political Action Conference Ronald Reagan Award. It must have been for her knowledge of Hollywood's version of the war, and not servicewomen's realities. Unfortunately, she is not alone in her narrow understanding of the debate. Only in February 2012, a Fox News contributor, Liz Trotta, attacked the Department of Defense for increasing spending on support programs for victims of sexual assault by blaming feminist activists for "wanting to be warriors and victims at the same time." Conservatives like Trotta seem to recognize the impact that new wars in Iraq and Afghanistan have had on changing front lines and women's role in the military, but they continue to rally against the financial "burden" and layers upon layers of bureaucracy "for women who are now being raped too much." The solution proposed is to simply remove the women, because their presence is an expensive distraction and the Pentagon is finally discovering that women and men are different and should not be mixing in the ranks. Trotta infamously proclaimed: "Now, what did they expect? These people are in close contact."[86]

Although weakened, DACOWITS has been focusing its energies on wellness (sexual assault and harassment) and assignments (removing all restrictions).[87] To compensate, a new organization, the Service Women's Action Network (SWAN), was established in 2007, to provide national policy advocacy and direct services to service women and veterans. In February 2012, a federal judge in New Haven, CT, heard their case against the Department of Defense and Veterans Affairs regarding the failure to provide SWAN with records documenting rape, sexual assault and harassment under the Freedom of Information Act. The same month, the Department of Defense announced that it would open an additional 14,000 combat-related jobs, mostly in the Army, or about 3 percent, leaving another 30 percent still restricted to men.[88] The biggest change is the lifting of a 1994 policy prohibiting women from jobs such as tank mechanic and field artillery radar

---

84 Donnelly (1995).

85 Book (2001).

86 Fox News, America's News HQ, February 12, 2012.

87 DACOWITS Annual Report 2011. Available at: http://dacowits.defense.gov/Reports/2011/Annual%20Report/dacowits2011report.pdf (accessed March 17, 2012).

88 Tilghman and Lance Bacon (2012).

operator that are carried out near combat units. This will open more than 13,000 Army jobs to women soldiers. A second change is an "exception to policy" that will allow the Army, Navy and Marines to open select positions at the battalion level in jobs women already occupy. The current policy, also set in 1994, prohibits women's assignments in intelligence, communications and logistics positions in units smaller than a brigade. This change will open an additional 1,200 assignments to women soldiers, sailors and Marines under the exceptions. Currently, 99 percent of all Air Force positions, officer and enlisted, are open to women. The figure is 66 percent for the Army, 68 percent for the Marines, and 88 percent for the Navy.[89] An additional 238,000 jobs remain closed to women.[90] In April 2012, at the request of Congress and then-Secretary of Defense Panetta, the Marine Corps launched a kind of pilot program to review the branch's policy on women serving in combat roles. This policy was to allow female Marines to work in areas that have been off limits to them, such as artillery, combat engineering, low-altitude air defense, amphibious assault and combat assault. Prior to the announcement of this new program, female Marines were not allowed to participate in the infantry training program. There are questions regarding this program that was supposed to open an Infantry Officers Course to 100 women in a one-year experiment. Two female Marines signed up and began training, but both dropped out of the 13-week program.[91]

Today, SWAN is the most effective organization advocating relentlessly on finding legislative solutions to issues facing both active duty service women and women veterans. They are currently working alongside a number of congressional members, including Rep. Loretta Sanchez (D-CA) and Senator Kirsten Gilibrand (D-NY), on getting the Gender Equality Combat Act passed, which would require a report on implementation of a termination of the ground combat exclusion policy for female members of the Armed Force. Moreover, SWAN has actively supported Senator Kirsten E. Gillibrand's Military Justice Improvement Act, which, if passed, would remove the prosecution of military sexual assaults from military commanders and hand it to criminal prosecutors.

Today, women constitute approximately 15 percent or roughly 281,743 out of 1.4 million active component military personnel, and comprise 7.25 percent of general/flag officers and 10.86 percent of the senior enlisted force. Although legislative progress has stalled, the participation of women in the military has expanded. In the years since the "risk rule" was rescinded, 1,000 women have served in Somalia, 1,200 in Haiti, 15,000 in Bosnia, and 8,000 in Kosovo, including women piloting combat aircraft.[92] More than 1.8 million women are American military veterans.

---

89   Parrish (2012).

90   *Ms. Magazine*, Feminist Wire Newsbriefs, May 24, 2012.

91   Burrelli (1998).

92   WREI Factsheet. Available at: http://www.wrei.org/Womenpercent20inpercent 20thepercent20Military/Womenpercent20inpercent20thepercent20Militarypercent20

On May 23, 2012, Command Sergeant Major Jane Baldwin and Colonel Ellen Haring filed a lawsuit in the U.S. District Court for the District of Columbia, charging that the military's ban on women in combat is unconstitutional and violates their equal protection rights under the Fifth Amendment. On November 27, 2012, SWAN and four servicewomen filed a lawsuit accusing the Department of Defense of violating their rights to equal protection under the law by continuing to maintain the combat exclusion rule. All four have served in Iraq and/or Afghanistan, two received Purple Hearts after being wounded in combat, two received medals in recognition of their combat service and one for engaging in direct combat after being wounded when her helicopter was shot down over Afghanistan. The lawsuit was filed in U.S. District Court in San Francisco and it charges that the DOD restriction based on gender is unconstitutional because it "categorically excludes women from certain combat positions, regardless of their individual qualifications and capacities." As such, these combat exclusions are not justified by a specific governmental objective, as the U.S. Supreme Court has required. The lawsuit specifically cites the practice of "attaching" women to combat units, yet female soldiers are left without the level of training provided to their male counterparts or the recognition that would enable them to advance.

For years, DACOWITS, WEAL, WREI, the NWLC and NOW have contributed greatly to the promotion of women's integration into the military and expansion of their role. Regardless of what administration was appointing women throughout the 1970s and 1980s, DACOWITS stood united with other women's groups outside the government seeking to expand the role women played in the United States military services. Together they changed the last bastion of machismo, and were instrumental in advancing women's equality in arms. Women's groups fighting for women in the military presented their cause tactfully to the American people in package that the population was already familiar with—equal opportunity. Despite the fact that DACOWITS has been deliberately dismantled, the fight for equality continues through a new generation of strong women's groups both inside and outside the Pentagon and Congress. Brian Mitchell was certainly correct in saying that "Were it not for intense political pressure, there would be virtually no women in the military."[93]

Italy

Women's participation in the economic, political and social life of Fascist Italy was largely limited to *Fasci Femminili*, the Fascist Party women's wing encouraging women's subordination, and occasional aristocratic and professional women's organizations, which handed over their household and childcare chores to female servants and housemaids from the lower classes. While the former possessed no political power, the latter organizations were allowed to operate on the condition

Chronologypercent20ofpercent20Legalpercent20Policy.pdf (accessed May 30, 2009).
   93   Mitchell (1998: xiii).

that they did not present a threat to party membership and allegiances. The most important women's professional association, FIDUS, or organization of women degree-holders, survived in the beginning although its calls to end labor discrimination were repeatedly ignored. But as its affiliation with the international women's movement, Jewish, lesbian and independent thinkers became more obvious, and as the regime created ANFAL, a carbon-copy Fascist version of FIDUS, the organization was asked to disband in May of 1935.

With the outbreak of World War II, things started to change as the regime abruptly abolished the policy of labor exclusion to fill jobs once held by men now being sent to the battlefield. While the government did not alter its emphasis on state's supremacy over family affairs and patriarchic and sexual hierarchy, women did manage to carve out their own destinies, even if temporarily. On the Fascist side, on April 14, 1944, Decree 447 was issued as the first official call for women to join the military. About 6,000 women joined the "Feminine Auxiliary Service" and some of them took up arms to fight the resistance even as it became clear that Mussolini's Republic of Salò was not going to survive.

However, most women joined the resistance. From 1943 until the end of the war in 1945, 70,000 women joined the Women's Groups for Defense and for the Assistance of Freedom Fighters (Gruppi di Difesa Della Donna e Per la Assistenza ai Combatenti Per la Libertà) and another 30,000 armed women joined the partisans to fight the Fascists.[94] Unfortunately, this large and active participation of women in the resistance movement did not lead to women's emancipation or gender integration in the military. They did not join to just fight the Fascist regime's chauvinist policies but to stop the regime that was sending their sons and husbands to their certain death. There was no feminist agenda as there had been no autonomous feminist movement since the Fascist Party came to power.[95] In fact, historians such as Alexander De Grand argued that "perhaps the greatest success of the [fascist] regime was the destruction of an independent political consciousness on the part of women."[96]

The Italian post-World War II feminist movement started to develop along with new political parties and realities of urbanization, the rise of the consumer society in the 1960s and the loss of women's jobs in agriculture that meant women increasingly worked at home within the patriarchal nuclear family. Although there were many different kinds of women's groups with large memberships, there was no united movement and the groups had no power as their goals and ideologies differed greatly. But any discussion of the Italian women's movement reveals two important subjects that are crucial in our understanding of their role and impact on gender integration policies in the military. The first, while there are significant differences between two major Italian women's groups and their political allies on notions of gender equality, their agreement and uniform rejection of the

---

94   Gori (2004: 72).
95   Wilson (1996: 84).
96   De Grand (1976: 947).

militarization of women's and men's lives delayed the process of gender integration in Italian Armed Forces. The second subject worth exploring concerns notions of difference and equality within the Italian feminist context and how these informed the perceptions of the state and military.

Among major women's groups were the Centro Italiano Femminile (CIF) and the Unione Donna Italiana (UDI). The CIF, with strong ties to the Catholic Church, focused its energies on the preservation of life and sanctity of family and marriage.

The only other group exerting influence was the UDI, the largest women's association, which was working with the Communist Party of Italy (PCI). The UDI was created in 1944–45 out of female defense groups, as some of the founding members were the wartime members of Women's Groups for Defense, such as Filomena Luciani and Leonilda Iotti. Holding the rank of major, Luciani fought for the resistance alongside American privates, while Iotti later became the first female Communist elected into the first Italian post-World War II legislature in 1946. With its headquarters in Rome, the UDI operated almost as any other political party given the structure and sophisticated agenda development. Moreover, unlike any other American women's movement, it encouraged its members to join a political party, trade unions and agricultural cooperatives. Having dual allegiance immediately set it apart as the movement could no longer claim autonomy, particularly as the membership was now represented in the legislative branch and with political agendas that no longer included only women's mobilization and promotion of women's interests. Members of both women's groups actively participated in politics, campaigned and backed party platforms, and supported different versions of gender relations. Early women's movements used the existing structure to organize themselves within these institutions simply because, "although mass movements have been successful in voicing and aggregating demands from various interest groups, they have not succeeded in reaching decision-making levels through bypassing party and institutional structures."[97] In fact, by the mid-1970s, feminists were a central component of many leftist organizations, and by the late 1980s old left parties such as the PCI established the principle of equal representation of men and women in their internal institutional bodies, Executive Boards and delegations, with a quota minimum of 33 percent of seats going to women.[98] Yet, some argue, these women's organizations never fully surrendered their autonomy to their political allies, as was the case in Eastern Europe and the Soviet Union, even though their political space, actions and agendas were often very constricted by the national and international political and economic realities of the Cold War (Pojman 2012).

The relationship between the UDI and the CIF was always cordial, but the differences between their definitions of gender roles and women's interests were largely based on the movement's ideological partner's understanding of women's and men's political and economic standing. While the CIF actively promoted the

---

97   Guadagnini (1994: 169).
98   Longo (2003: 8).

Catholic Church's version of gender, and hence recommended more feminine jobs to be done by women, and masculine jobs by men. It never discouraged women's participation in politics and economics, but identified women's feminine attributes as supporting that of the breadwinning masculine men to get through tough times. It also saw Italian friendship with the United States and its membership in NATO as the way out of economic troubles plaguing the post-war period. Their understanding of "women's equality was through their complementarity with men."[99]

The UDI's love affair with Joseph Stalin's materialism, and supposed gender equality in the Soviet Union greatly influenced the way gender was constructed by the movement. The return of women into the domestic sphere for the sake of the nation's welfare and their manly breadwinners brought back memories of Fascism. It is important to emphasize that although the war was long over, the battle against Fascism was not as the ideology remained relevant in political discourse in various other parts of Europe, not to mention in Italy's political arena. The awareness of its presence had a major impact on the mobilization of the women of the UDI, their choice of allies, and their political and economic vision for the women of Italy. Moreover, they positioned themselves against capitalism as it represented the exploitation of both women and men and eliminated social services that facilitated equal participation of both in socioeconomic and political life. Their opposition to NATO as another form of militarization and return to patriarchal systems of domination was directly linked to their opposition to capitalism. Both NATO and economic liberalism were antithetical to gender equality, as they not only corrupted women but also men's lives.

The difference between the two associations became more acute with time, particularly in the 1960s as the issues of class, religion, marriage and labor entered the legislative arena. Discussions of the military as a profession were not part of these conversations, but the fear of militarization of women's and men's lives was a shared concern of both women's associations. The UDI's thinking was largely tied to the PCI's antagonistic understanding of the Italian Armed Forces as the conservative, anti-Communist institution encapsulating the dangers of an imperialist war. Military preparations by NATO, which included further militarization of the Italian population, were to be hindered and resisted. The Catholic Church has traditionally rejected the notion of women's involvement in the military and conflict, because they were not seen as appropriate places for feminine women. Likewise, the women of the CIF did not spend their energy on initiatives concerning gender integration as they found the subject contrary to pacifist discourse and the teachings of the Church.

This history of socialist politics behind the feminist movement played a significant role in the policy issue of gender integration in the military. The idea was actively opposed by virtually all women's organizations. The inclusion of women in the Italian armed services was first brought up in 1963, when Law

---

99   Pojmann (2013: 72).

No. 66 abrogated the 1919 law and 1920 regulations and gave women access to all public professions and governmental jobs, with the possibility of openings in the military, too. But since the women's movement was not interested in participating in the military, it did not participate in what it regarded as another attempt by conservative right-wing parties to co-opt and corrupt women for their capitalist exploitation. Parliament rejected an amendment proposed by the government aimed at excluding women from employment in the armed forces, but it left the regulation on women's service in the Army and Special Forces to future particular laws.[100] Yet, the views of the PCI remain debatable as Togliatti, its political leader, was quoted as saying that eventually women could fill those roles and functions that were "not appropriate for men."

The UDI continued to mobilize with the old left wing, against the CIF on the anti-divorce referendum led by the Christian Democratic Party in 1974 and the rejection of the 1976 legalization of abortion bill by both the Christian Democratic Party and the Neo-Fascist Party.[101] As student protests swept across the European continent in the late 1960s, around the country, issues and mobilization regarding gender started to change. The extra-parliamentary leftist (*sinistra extraparlamentare*) groups, parties and movements calling for direct political action and pointing to the involvement and participation of the masses without intermediaries, were attracting young women. These women, unhappy with the political space given to them within the existing political structures, were interested in developing a process of political change outside those oppressive structures. By now joining consciousness-raising (*autocoscienza*) organizations and starting their own centers and magazines to understand women's oppression, these groups formed and began various initiatives. Their discussions in assemblies and meetings quickly distinguished liberation from emancipation, the former dealing with the radical transformation of everyday life while the latter was seen as having a more limited focus on public life, including the work place (Cavarero 1987, 1999). This new radical version of gender relations in an Italian context was largely brought on by the new Movimento Di Liberazione di Donna (MLD), associated with the Radical Party. Unlike the women's organizations of the old left, such as the UDI, which had concentrated their energies on opening institutions to the female vote, and firmly believed in parliamentary democracy, the movement that developed in the early 1970s was permeated by a deep aversion to the state. The MDI interpreted state's notion of equality as just another way to subject women even further. Equality of opportunity meant emancipation and becoming more like men and that was seen as an imposition of asexual identity that turned women into just another political category developed by patriarchal, right-wing governments. Instead, they

---

100   Legge n.66 del 9 febbraio 1963. Ammissione della donna ai pubblici ufficie dale professioni. http://www.donne-lavoro.bz.it/download/284d2588_v1.pdf (accessed June 5, 2009).

101   Guadagnini (1995: 153).

wanted revolutionary change.[102] That change was not going to take place under the Christian Democratic Party regime, because as some would argue, "Italian women continued to live as a minority group in a situation of serious inferiority, not unlike that reserved for them by the Fascist regime that had just passed."[103]

The issue of gender in the military continued to be rejected as the movement:

> opposed authority and institutions ... and stressed the need for a distinct and separate identity for women as women within institutions ... With "difference" as its trademark, the movement was more interested in reproduction and sexual issues than in labor issues, and it did not try to achieve positions of political and social power. Rather than seeking to become a pressure group, the movement in the 1970s stood aloof from decisional mechanisms.[104]

What the 1970s brought on was the feminist challenge of social institutions in which dominant "patriarchal" and "masculine" values led to the supremacy of man in Italian society. The new feminist movement expressed its strong opposition to this "macho" organization of society as it represented the imposition of a hierarchy typical of the male world, and also criticized "femininity" as an imposition of traditional culture and mass media. Military services were the ultimate expression of state's institutionalization of patriarchal hierarchy and oppression, and much like Fascism, was abhorred.

Unlike in the United States, there was no uniform front fighting for "equal rights" and there were no women's groups trying to fight for equal position and obligations in the military services. Terms of reference and understanding of what gender equality brought, both to the society and military, were in stark opposition to the debates taking place in the United States. That only started to change after the collapse of Communism, when the changes in the Italian political environment and international security context prompted a reexamination of the civil military relations, as well as the structure, role and operational capabilities of the Italian Armed Forces.

By the time the Cold War neared its end, the Italian military was composed of 385,100 members, of whom 258,000 were conscripts. By the early 1990s, Italy was one of the few states in the world sending conscripts abroad, on peacekeeping missions in Somalia and Mozambique. They ended their missions in 1995, having deployed only volunteer troops to Bosnia.[105] As the ethnic wars continued to rage on the Balkan peninsula through the 1990s, it was becoming clear that Italian forces were not modern or sophisticated enough to participate in peacekeeping operations, and parliamentary debate was starting to focus on a transition to either a mixed model or volunteer force military model.

---

102   Yasmine Ergas (1982: 262).
103   De Clementi (2002: 333).
104   Guadagnini (1995: 152).
105   Villani (2005: 382).

In 1992, the Italian government decided to conduct an experiment near the Sabatini Barracks, home of the 8th Regiment "Lancers of Montebello" in Rome by inviting women to join the army ranks for a day and a half. Thousands of Italian women responded to the state's call and applied to the Ministry of Defense, demonstrating that things were starting to change in Italy. Twenty-nine Italian women were chosen to participate and were given a chance to live in the barracks for 36 hours, training for and carrying out military activities. This was a test that was to hopefully provide the Ministry of Defense and military services with more information to help them better formulate proposals for the new defense model law which was to be voted on in parliament. At that point, volunteers made up only 4.9 percent of the entire force, and the reform was to add 65,000 more.[106] The woman soldier was to be a part of the new grand defense personnel model that would include both conscription and voluntary systems.

Three years later, on May 25, 1995, 12 of those 29 young women in the experiment formed an organization called ANANDOS (National Association of Women Soldier Aspirants). These women started a very serious debate that involved radical feminists, the Catholic Church, and the new left. But as radical feminists and women in peace movements did not want to take part in pushing for what they called "equal opportunity to kill," the legislative battle was left up to the quickly rising ANANDOS membership. This new organization mobilized nationally to pressure the government to give their members full equality in enrolling in military service. In a survey conducted on June 15, 1999, 600 ANANDOS members were interviewed to find out which positions in the military they would like. Forty-three percent of the women declared a desire to enter military academies and become officers; 22 percent wanted to be non-commissioned officers, and 12 percent simply said they would like to join the armed forces. Of these 12 percent, 21 percent wanted to join the Air Force, 19 percent to serve in the Army, 10 percent in the Navy, 14 percent in the gendarmerie, and 9 percent sought to join the financial police.[107]

Such surveys helped ANANDOS raise awareness and lobby parliamentarians unfamiliar with their cause. The group started to promote the recruitment of women in the armed forces through the media, online forums and websites, and subsequently collaborated on the writing of the text of the actual law.[108] The bill was introduced on January 15, 1997, by the Social Democratic parliamentarian Valdo Spini, a member of the Italian Chamber of Deputies and the president of Chamber's

106   "Oggi Trenta Donne Soldato." *Corriere della Sera*, November 23, 1992, 12.

107   Carabinieri (military and civilian police) and Guardia di Finanza (financial police) are both considered branches of the Italian Armed Forces. Ministero della Difesa Archivio Servizio Femminile (Ministry of Defense Archives on Female Service). Available at: http://www.difesa.it/Approfondimenti/ArchivioApprofondimenti/Servizio+femminile (accessed June 10, 2009); Military Police or Carabinieri and Guardia di Finanza.

108   Italy National Report to the Committee on Women in NATO 1999–2000, 1. Available at: http://www.nato.int/ims/1999/win/report99.pdf (accessed June 15, 2009).

Defense Commission.[109] As part of the inquiry on reform of conscription and new military needs, on February 26, 1997, the Defense Commission held a hearing with the National Commission for Equality and Equal Opportunity leadership, and Deborah Corbi, the president of ANANDOS. She supported Spini's proposal for the creation of a voluntary military service in which women would be equal to men, and rejected suggestions of the creation of an exclusively female auxiliary corps, because such an arrangement could relegate women to subordinate and supporting position within the armed forces. Assigning tasks to be on the basis of sex seems, according to Corbi, a violation of the Italian Constitution, which does not speak of distinctions between men and women but underlines that the defense of Italy is the "sacred duty of every citizen."[110]

ANANDOS was not fighting for difference, but for the same rights, the same jobs, the same opportunities as men, and to be part of the same military hierarchy. In her proposal, Corbi suggests that all citizens should be able to enlist, carry out military jobs, have access to leadership roles (academies for officers, non-commissioned officers and other graduate programs) and consequently, based on merit, achieve higher rank without distinction of gender, marital status or number of children. Corbi and the other women of ANANDOS did not support quotas and guaranteed seats in any department, only requesting that the policymakers grant them equal pay, the same benefits, same pension, and same representation. The only exception was maternity, which they regarded as "natural" distinction, and hence women ought to be afforded the same rights under the law in force for women performing ordinary jobs, with a reduction in work effort from the early months of pregnancy.[111] Present at the same hearing was Maria Celeste Nardini, a member of the Communist Refoundation Party (Partito Rifondazione Comunista—PRC), which was formed after the PCI disintegrated in 1991. Her explicit disagreement with the proposal was based on the old Communist and radical militant feminist argument—that of gender difference. Instead, she demanded greater reflection and debate on the question of opening the military to women given the rise of conscientious objection applications, and an increase in discomfort among men who perform military service.

Only a couple of weeks later, on March 8, International Women's Day, Corbi and ANANDOS and a group of influential Italian female politicians and thinkers from both the center-left and right released a statement titled "Please not call us anymore 'soldiers in a skirt.'" In it they reminded the public and media that Italy was the only European state that does not allow women access to the military

---

109   Commissions are equivalent to American Congressional Committees.

110   Italian Constitution, Article 52. "La *difesa della Patria è sacro dovere del cittadino.*"

111   XVIII Legislature, IV Commissione Difesa Indagine Conoscitiva Riforma Della Leva E Nuovo Strumento Militare, mercoledì 26 Febbraio 1997. Avaiable at: http://www.camera.it/_dati/leg13/lavori/stencomm/04/indag/riforma_della_leva/1997/0226/s000r.htm (accessed April 2013).

force. The statement was a result of the media's reference to ANANDOS women as "soldiers in skirts" and demanded not only the immediate opening of the Italian armed forces to all, but that women soldiers be treated as simply soldiers in all respects.[112]

They were heard. The final proposal was examined by the lower House on July 24, 1998, and approved seven days later. Anna Finocchiaro, the Minister for Equal Opportunity, immediately commented that this law, if passed, would eliminate the final obstacle for women in public administration, while Spini argued that "the approval comes at the moment when the Italian Armed Forces are engaged in many peace missions abroad ... and we were the last country of NATO to allow women in the government." Italy was now an active participant in different types of military operations, largely multilateral and peacekeeping operations including in Albania, Croatia and Bosnia. By repacking, and explaining Italian military engagements within a pacifist frame, the government was also influencing public opinion and softening the negative memories of a Fascist military profession as offensive, aggressive, and masculine (Iganzi, Giacomello and Coticchia 2012). Such discourse served to present a more positive image and justify the passage of gender integration policies, even if the reality was inconsistent with such representation of both types of operations and change in military masculinity.

On September 16, 1998, the bill was assigned to the Senate and examined until July 15, 1999. With a number of modifications, it was finally approved on July 21, 1999.[113] In the end, the bill was returned to the lower House, which approved it on September 29, 1999, with an overwhelming majority, with 592 votes for, nine against, and nine abstentions. All nine votes against the bill were from the PRC, and the nine abstentions came from the Green Party (Verdi) and the Party of Italian Communists (Partito dei Comunisti Italiani PdCI). With this last approval, Law No. 380 was finally signed on October 20, 1999, allowing women to enter military academies beginning in the year 2000. In 2001, they would be allowed to enter as non-commissioned officers, and in 2002, as enlisted soldiers. Although there were no obstacles to women's career advancement, they were not allowed to be assigned in situations of extreme risk of direct contact with the enemy

The Italian National Report to the Committee on Women in NATO proudly declared that the notion of the woman soldier was borne from the demands of civil society, namely the women's organization ANANDOS. What it fails to mentions is that one of the main reasons why this women's organization became so strong is that the old left was no longer a significant player in the Italian political arena and that radical feminists were no longer very relevant.[114]

---

112   8 Marzo Donne Soldato Non Chiamateci Militari In Gonnella.

113   "Arrivano le donne soldato ma non in prima linea." *La Repubblica*, September 29, 1999.

114   Italy National Report to the Committee on the Women in the NATO 1999–2000, 1.

After the passage of this law, the autonomous social movements felt betrayed by the willingness of new left parties such as the Social Democrats to work with parties they labeled Fascist and imperialist. Those movements still argue that man-woman parity had nothing to do with this legislation, and that it exploits the talents of women on the organizational and logistical level far away from the combat. According to these old left radical parties, this law is an attempt by the imperialist army to appear more democratic and popular, because the real change would be if the Army served only to defend the country and promote peace, not invade other countries. While it is difficult to defend their radical points of view, it is interesting to note that Corbi is considered one of the "donne di Fini," or women of Fini, the leader of the Alleanza Nazionale (AN), the 1990s political party reincarnation of the neo-Fascist MSI-DN.[115] Clearly, the women of the "third wave" are no longer inclined to follow the leftist class struggle rhetoric.

On January 11, 2000, the European Court of Justice decided that limiting women's role in the military was a clear "violation of the principle of equal opportunity ... in about half of all European armed forces, women are admitted to combat units without restrictions"[116] The 20 percent quota for female recruitment (percentage of the total available places) was abolished in 2005, followed by the elimination of employment restrictions in 2006.[117] With the exception of submariners and Special Forces, all military jobs are open to women today. In 2008, there were 10,000 women serving with the Italian armed forces[118] and since 2001 they have been deployed to Kosovo, Afghanistan, Iraq and Lebanon.[119]

The reforms regarding gender integration coincided with passage in the Italian parliament of the Armed Forces Reform Act in November 2000 to make the transition from conscription to an all-voluntary army by 2005. The transition was viewed by policymakers as necessary to reduce costs and to meet the challenges facing Italy's armed forces, including participation in the European Defense Forces, peacekeeping missions, and the defense of interests beyond the national borders (read Iraq and Afghanistan). The decision to move to an all-volunteer force was matched with a decision among Italian policymakers to preserve and expand significantly the civilian service program which had been developed for conscientious objectors. By 2000, Italy was the only NATO state that not only did not have women serving in its military, but that also had very few professionals with university diplomas and expertise. Within this new organization of the armed services, the presence of qualified women was certainly seen as a positive addition and the Italian parliament was more than willing to accommodate ANANDOS to help fill the ranks with eager young women with necessary qualifications.

---

115   Latella (2002: 9).
116   Haltiner and Klein (2005: 22–23).
117   Italy National Report to the Committee on the Women in the NATO 2008, 2.
118   Italy National Report to the Committee on the Women in the NATO 2008, 1.
119   Italy National Report to the Committee on the Women in the NATO 2008, 3.

After almost 10 years, in 2004 ANANDOS dissolved and was replaced by the Gruppo Donna Soldato (Group of Women Soldiers), which today is the only organization in Italy dealing with women soldiers and young women aspiring to serve. This new organization took over ANANDOS's agenda and their friendly relationship with the Ministry of Defense. Their biggest battle today is eliminating current height (161 cm) and age (26 years old at the time of entry) restrictions, but the Ministry of Defense Communications and Research Representative Captain Rosa Vinciguerra in a recent radio show argues that this is not unjust, because men also have height and age limitations. However, she does acknowledge that not all is "roses and flowers when it comes to the four services of the Armed Forces, but there is a march ahead."[120]

In an interview in February 2008, Sabrina Bretoni Piazza, the president of Gruppo Donna Soldato (GDS), acknowledges problems of hazing and sexual harassment but surprisingly identifies the exclusion of the women and men of the Italian military from Law 151 of 2001 as the biggest problem women face. This law allows public administration workers with children younger than 3 to be transferred temporarily to the region or province where the other parent resides. Basically, women and men are forced to leave their small children at home, which, according to the GDS president, is contrary even to the constitutional protections of the family as a sacred unit. She agrees that women are still unable to say that they are on an equal footing with men, and suggests that it is necessary to have an equal opportunity commission within the armed forces to deal with it.[121] This concern was also voiced in the aforementioned radio program by Dr. Donatella Linguiti, the Under Secretary for Rights and Equal Opportunity, and First Marshall and military union Cocer delegate Luca Tartaglione, who argue that there is no reason why the military should not have a commission for equal opportunity when all other branches of public administration already have one. It is simply deemed an unnecessary waste of money.

Today Italian women serve in Afghanistan. In Herat there are about 150 women, with the same operational duties as their male counterparts: they command platoons, pilot helicopters, serve as riflewomen or engineers, engage in the task of mine and explosives deactivation. They also serve as doctors, nurses and psychologists. Others choose to serve with the components of the FET (Female Engagement Team), whose task is to interact with Afghan women to improve their conditions. In a recent newspaper interview, Italian women claim that being a woman in war does not take away from their femininity, and their duties

---

120   Program "Donne in Forze Armate" (Women in the Armed Forces) on Italian state radio station RAI on January 3, 2008. The program is available at: http://www.radio.rai.it/radio1/laradioneparla/view.cfm?NOTIZIA=237133&DATATEMA=2008-01-03 (accessed June 15th, 2009).

121   Giuseppe Paradiso, "La condizionedella donna nelle forze armate" (Condition of women in the armed forces), *GrNet* the information service for those in Defense and Security, February 9, 2008, 1.

as mothers and wives. "We are military but we are still women," one of them explains. "But in the end, we are no different from our male colleagues because we are all away from our families and suffer equally. We Italians are not very accustomed to parents who leave children for long periods. There is no difference: the attachment to a child depends neither on sex nor on the distance."[122] Yet, throughout the article, female soldiers keep emphasizing their femininity and different approach to problem-solving and military work as a way of showing their unique contribution to the institution, as if they want to reassure the country and Italian men, that Italian women are and always will be first and foremost mothers and caretakers, and soldiers second. The call for integration in Italy was driven by a group of very eager women who wanted to be seen as soldiers first, but today there are many questions about the independence of Italian servicewomen groups. Italian researchers have expressed their frustration privately to me and others who have wondered publicly on the GDS online forum whether obtaining information about Italian female soldiers is a utopian dream. Some researchers have had their questionnaires even edited by the Ministry of Defense, and others were unable to find a single woman willing to anonymously discuss their position, problems and successes within the ranks. In fact, one such request for information about the reasons that prompted women to enlist and their struggles in the theater of operations, was turned down by Bretoni Piazza with the following sentence: "Disgusted. Not all that glitters is gold." Most of those researchers were frustrated by some sort of code of silence regulating the behavior of these soldiers who are unwilling to participate and comment even anonymously. Although I have asked them to provide information regarding total percentages of women in the Italian armed forces, as well as percentages of commissioned and non-commissioned officers, I was told by Bretoni Piazza to submit my request to the armed services. It is surprising that the main organization representing women in the military would not have such information readily available or even attempt to show any interest in obtaining it. Questions regarding political parties or any other women's groups that have provided legislative support for women in the military were answered with one simple line: Ministry for Equal Opportunities. No political party, just the ministry. Bretoni Piazza has always insisted that her organization is an independent initiative which aims to ensure the needs and aspirations of Italian women and therefore is not part of an ideological battle. Yet, it is possible to draw some conclusions despite this silence. The narrative shows that the vociferous left-leaning women's movements of the 1960s and 1970s that perceived the state as a potential threat and the military as just another way of corrupting women did not participate or work on legislation that would open the Italian military to women. But the silence also speaks volumes. It shows that the new generation of women's

---

122   "Herat, l'8 marzo delle donne soldato -Per noi un giorno come gli altri." *Corriere Della Sera*, March 8, 2012. Available at: http://www.corriere.it/cronache/12_marzo_06/donne-soldato-festa-herat_17f091ce-67a2–11e1–894d-3b3e16fcb429.shtml (accessed October 10, 2012).

groups that has been more interested in working with state officials and political parties has been successful in promoting equal opportunity in arms. While it might be seen as the realization of all left-leaning movements' fears regarding the cooptation and corruption of women's lives, Bretoni Piazza insists on asserting her "natural right" to build a better future by including more women in Italian military.

Gender Equality in Workplace

USA

Women's groups fighting for gender integration in the military presented their cause tactfully to the American people in the package that the population was already familiar with—equal opportunity. Mary Fainsod Katzenstein has argued that the:

> emergence of organized feminist activism would not have been possible without the law's legitimation of equal opportunity ... the courts and congressional endorsement of antidiscrimination norms gave advocates of equality a voice that otherwise might well have been overpowered within the institution.[123]

In fact, since World War II the progress of women in the military has always following the progress of women in the market place. By the time that war ended, 19 million women, or 36 percent of the U.S. work force, were employed in the American war effort. Although Rosie the Riveter and millions of her fellow women helped keep the American economy running, as World War II ended and men returned from the front lines, women left the factory to retreat into the home, just as female soldiers did. Labor Committee Chairwoman Mary T. Norton urged women not to do so. She understood the pressure on women from industry and labor unions to vacate jobs that their husbands and brothers were seeking to obtain:

> This is the time for women everywhere to prove that they appreciate the responsibility they have been given. Women can't be Sitting-Room Sarahs or Kitchen Katies. They have homes to keep up, food to prepare, families to clothe ... but they have their world to make ... American women today stand on the threshold of a glorious future ... They can grasp it ... or they can let it slide. Women are going to be pushed into a corner, and very soon at that.[124]

Although some women did return home by the time the Women's Armed Services Integration Act was passed in 1948, the percentage of females in the labor force

---

123   Katzenstein (1998: 80).
124   U.S. House Document (2006: 144).

dropped to only 3 percent. From that low point, over the years that number continued to rise.

The passage of the 1964 Civil Rights Act gave women another opportunity by eliminating all discrimination in the labor market based on sex. As discussed above, this Act created a momentum which allowed women's movements to make greater demands seeking expansion of women's participation in the military. It allowed the movement to situate their cause within a larger context of discrimination of women in the work place and present it to both the public and Congress as such.

In 1973, as the conscription system ended and an All-Volunteer Force was created, the Pentagon realized that it would need to recruit fewer, but more qualified and skilled, professionals to fill its jobs. But in 1978, an average of 60 percent of male Army entrants possessed a high school diploma compared to 81 percent among 19–20 in the general population, and 71 percent of draftees in 1964. In addition, in 1964 over 40,000 persons with some college education entered the Army's enlisted ranks; in 1978 the figure was less than 5,000.[125] Therefore, the Department of Defense was forced to tap into the female labor market and recruit the most qualified to fill vacant positions.

In 1991, Congress passed the Non-Traditional Employment for Women Act, requiring training and placement of women in fields traditionally dominated by men, and the Civil Rights Act of 1991, which provided for punitive damages in cases of intentional discrimination.[126] This also marked the beginning of changes in the military ranks, as there were no more legal obstacles to women in other civilian labor sectors that were previously dominated by men.

In 2005 Janet Hoffheins, deputy director of DOD Civilian Personnel Management Service, argued that the modern military force is, and will continue to be, composed of highly competent and dedicated women. She continued by saying that "as we move forward into the 21st century, our challenge is to ensure that the department attracts and retains the best and brightest ... the right people with the right skills to achieve the mission." Hoffheins used the Defense Manpower Data Center and the U.S. Census Bureau reports to show that the number of active-duty women officers increased in several non-traditional occupations, such as engineering and maintenance, tactical operations, and supply and procurement, whereas the number of enlisted women increased in the areas of tactical operation and supply and procurement. The top five occupations in 2004 for active-duty women officers were nurses, physicians, biomedical sciences and allied health officers, health services administration officers, and manpower and personnel. At the same time, the top five occupations for active-duty enlisted women were

---

125   Moskos (1979).

126   Civil Rights Act of 1991 (Pub. L. 102–166), as enacted on November 21, 1991 and published by the EEOC. Available at: http://www.eeoc.gov/policy/cra91.html (accessed May 23, 2009).

general administration, supply administration, general personnel, general medical care and treatment, and operators and analysts.[127]

Today 68 percent of American women aged 15–64 are employed and women make up 47.5 percent of total American work force.[128] Of all jobs in scientific research and development in 2007, U.S. women held 43 percent and 34 percent in national security and international relations.[129] Women in the United States make up 55 percent of professional and technical employees today.[130] In certain professions, particularly high-paying management and professional occupations, women are already at 51 percent. Men have also been outnumbered by women in occupations such as financial managers, human resource managers, biological scientists, writers, accountants and budget analysts, educational administrators and medical and health services managers.[131] These are the same occupations that women in the military are largely occupying today.

Italy

Italy has been a latecomer when it comes to equality of women in the economic sector. This section demonstrates that the gender inclusiveness index remains low due to continued low levels of female economic activity and low numbers of women with technical skills necessary to fill military positions.

In the aftermath of the unification of the Italian peninsula and prior to the arrival of Fascism, women were largely considered "the charming, empty-minded, useless creatures"[132] of lower level whose primary job was to take care of their families. Most of the women working in professional fields were teachers and school principals. In other fields, such as law, business and even medicine, women were scorned and quickly excluded by their male colleagues.[133] The majority of Italian women worked in agriculture, and failed to receive any professional or technical training. This was particularly true in southern Italy, where family

---

127   Williams (2005).

128   World Bank GenderStats Data Query for data from 2008. Available at: http://www.genderstats.worldbank.com (accessed May 23, 2009).

129   U.S. Department of Labor, Bureau of Labor Statistics, *Women in Labor Force: A Databook*. December 2008, Table 14 Employed Persons by detailed industry and sex, 2007 annual averages. Available at: http://www.bls.gov/cps/wlf-table14–2008.pdf (accessed May 23, 2009).

130   United Nations Development Program, *Human Development Indices 2008*, Table 5, 41–44. Available at: http://hdr.undp.org/en/media/HDI_2008_EN_Tables.pdf (accessed on May 23, 2009).

131   U.S. Department of Labor, Bureau of Labor Statistics, *Employment and Earnings, 2008 Annual Averages and the Monthly Labor Review*, November 2007. Available at: http://www.dol.gov/wb/stats/main.htm (accessed May 23, 2009).

132   Zampini Salazar (1984: 158).

133   Zampini Salazar (1984: 162–163).

ties and community came before a woman's professional and educational advancement. In fact, women were viewed more as a family possession tied to the family home and always supervised by the kin rather than persons capable of making any independent decisions. Their primary responsibility was their family household and, at most, farm tasks, and some clothing and food production. Women working outside their nuclear family homes were seen as undesirable and immoral potential marriage partners.[134]

From Italian unification until the 1920s, the schooling of women was also not a priority, particularly among the poorer classes, where women's labor from an early age was necessary for family survival.[135] Among the wealthier classes, girls attended school at a higher rate, but educational associations and organizations trying to expand opportunities were largely limited in their aim and scope, and were often despised by true women's rights pioneers such as Fanny Zampini Salazar, who argued that:

> nothing could better reveal the subjection of our women to prejudices and old ideas than this association of theirs, which pretends to promote woman's culture by a weekly lecture, mostly regarding ancient history, and carefully excluding any and all of the modern questions regarding social, educational, legal or political matters.[136]

The arrival of Fascism in the 1920s brought with it the *Nuova Italiana*, or New Italian Woman. Victoria De Grazia superbly captures the gender dynamics from 1922 until 1945 in one single sentence: "Mussolini's regime stood for returning women to home and hearth, restoring patriarchal authority, and confining female destiny to bearing babies."[137] Italian women were to become the main protagonists of Mussolini's pronatalist policies. During this period women were not only supervised by their kin but also by the state, which used them to fulfill Fascist, imperialist and mercantilist agendas. Besides deliberately limiting women's economic activity by preventing them from getting or staying in jobs, Mussolini's regime also nationalized brothels and banned contraceptives, abortion and sex education. By doing so, De Grazia argues, the Fascist and militant state "sought to nationalize Italian women."[138]

After World War II ended and the new Italian Constitution was written in 1948, Article 3 guaranteed equal social status and equality before the law, without regard to sex. In terms of labor, Article 37 on Equality of Women at Work reads:

---

134   M. Cohen (1993: 19).
135   M. Cohen (1993: 22).
136   Zampini Salazar (1984: 161).
137   De Grazia (1993: 1).
138   De Grazia (1993: 6).

Working women are entitled to equal rights and, for comparable jobs, equal pay as men. Working conditions have to be such as to allow women to fulfill their essential family duties and ensure an adequate protection of mothers and children.[139]

In defining women's equality, the constitution discriminates by assigning the family duties and childcare explicitly to women. Women's employment should not get in their way of performing primary duties as mothers and wives. With such an article being part of the constitution, it is no wonder that post-war modernization, although largely characterized by the transition from village-based extended families to urban nuclear families, did not produce greater numbers of women in the work force. In fact not much was done to ensure equality between women and men. By 1963, women had only gained access to public and administrative offices, but another 14 years would pass before the 1977 Law No. 90 finally gave women equal opportunity in the work place.[140] It was only after this law was enacted that women started to enter the labor market in higher numbers, and between 1970 and 2000 female labor participation increased by 70 percent while males' participation was static.[141]

Italy was also slow in establishing agencies that would implement equal opportunity and promote gender equality. Almost a decade after the majority of Western European states established official state agencies to accomplish those tasks, in 1983 the Italian government finally decided to create the Equal Status Committee (ESC) attached to the Ministry of Labor and Social Security for the sake of implementation of equal treatment and equal opportunities for men and women in the work place. A year later, another agency was established within the prime minister's office, called the Equal Status and Equal Opportunity National Commission (ESNC). With this agency the government sought to "… further women's active participation in political, social, and economic life, especially in the economic and political decision making process."[142]

Guadagnini argues that the passage of the 1977 law and the establishment of both the ESC and ESNC were not a result of domestic political pressures or women's groups organizing to lobby, but rather resulted from international pressure to meet the directives of the European Community regarding equal treatment and pay, and pressures from the UN to expand opportunities during the International Decade

---

139   La Costituzione Italiana (Italian Constitution) available on the website of the President of the Italian Republic. Available at: http://www.quirinale.it/costituzione/costituzione.htm (accessed May 26, 2009).

140   Legge n. 903 del 9 dicember 1977, Parita' di trattamento tra uomini e donne in material del lavoro. Text available in Italian at: http://www.donne-lavoro.bz.it/291d2678.html (accessed June 5, 2009).

141   Albanese (2006: 149).

142   Guadagnini (1995: 156).

for Women.[143] As such, they lacked contacts with the grassroots organizations and women's groups, and therefore did little to promote feminist goals of breaking away from the patriarchic understanding of women's primary duties to their families.[144]

Today, only 52 percent of Italian women are employed, a very low figure compared to the other OECD states, whose average is 63 percent,[145] and 5 percent of top management positions are held by women.[146] According to a report published in April 2007 by the Ministry of Labor and Social Policy, 54 percent of "precarious workers" are women. Although they have obtained higher levels of education than men, 20 percent of women in Italy are employed in positions requiring less education. In fact, the numbers are so dismal that only 1.2 percent of women accumulate a total of 40 years of social security contributions, and 52 percent less than 20 years.[147] Similarly, IRIS research shows that women get 80 minutes less leisure time than Italian men, the biggest differentiating factor in comparisons with the other OECD states. The researchers suggest that an Italian woman's family remains her main responsibility.

But there is another argument by old feminists, social researchers and commentators why Italy continues to lag behind other Western European democratic states. They argue that feminism and the fight for equal opportunity of the 1970s are out, and show girls and sex symbols are in. Italy's current Minister of Equal Opportunities is Mara Carfagna, a former show girl and protégé of Prime Minister Silvio Berlusconi, whose media empire makes billions from TV shows full of women's flesh and who himself enjoys the company of high-end prostitutes.[148] Whether the women of Italy can really expect their situation to change in the near future with these two at the helm is beyond the scope of this study, yet it is a necessary point to consider when studying equal opportunities in both civilian and military labor sectors.

This dismal situation of women in the labor market in general and in professional and technical fields reflects the situation of women in the Italian military. As in the case of the United Stated, the two sectors seem to be mirror images. In neither are there legal barriers to the women's employment of women, but in both sectors their numerical presence and the quality of that presence remain low. Although women have been recruited for the last eight years, they are still lacking in numbers of both enlisted soldiers and junior and senior officers. Women make up only 2.6 percent of the entire force, and represent only 0.2 percent of senior officers, and 1.1 percent of junior officers. Unless the situation changes regarding the employment of women and their percentages in professional and

143   Guadagnini (1995: 155).

144   Guadagnini (1995: 165–166).

145   OECD Labor Force Statistics by sex and age indicators. Available at: http://stats. oecd.org/index.aspx (accessed May 31, 2009).

146   Wallis (2008).

147   Santi (2007).

148   Poggioli (2008).

technical fields in civilian society, we should not expect a greater inclusiveness of gender in the Italian military service.

Conclusion

One cannot consider gender integration in the military devoid of its historical context and political actors seeking to use men and women to achieve security agendas. The cases of the United States and Italy illustrate that although quantitative analysis correctly identifies the women's movements and gender equality in the workplace as key policy determinants, process tracing reveals the prevalent societal gender narratives and their impact on both women's movements and gender equality in the workplace. During World War II American government policy and propaganda were aimed at turning every American woman into muscular and masculine Rosa the Riveter, while Italian Fascist government underlined women's feminine characteristics and sought to limit their role to that of mother and homemaker, by encouraging open labor discrimination in favor of procreation. It is within this context that two different types of women's movements were born, two different workplaces were created and two different levels of gender integration policies in the military. While women's military engagement and achievements in the United States were part of official government policy, Italian women engaged in war were largely battling their own government. Equally important were political coalitions formed in the post-World War II period that defined legislative discussion in years to come. While women's movements in the United States supported equal service as a matter of equal employment and citizenship, Italian difference feminists saw it as corruption of women's bodies by a patriarchic, sexist and Fascist institution. Official disinterest of Italian women's groups in gender integration in the military fearing corruption and militarization of the women's agenda by right-wing political parties stalled the reform and left the issue unresolved for much of the twentieth century; unlike the American women's movement that perceived gender relations in terms of structural oppression and subordination that cannot be solved by means of legal battle. While the Italian feminist movement became a separatist movement, not fighting *for* equality but *emphasizing* difference and class struggle, the American women's movement was working within a wide nexus of activists, militarists and policymakers, and articulating their demands regarding gender integration in the military within a larger equal opportunity and anti-discrimination framework. Italian feminism set itself apart both with its agenda and allies, but despite their solidarity and agreement that industrialist, capitalist and misogynist men and media are the problem, there were no single voice solutions. Their other issue that hampered gender integration was the inability of this separatist movement to understand young women, who were genuinely interested in joining the armed forces and changing political discourse regarding the role of military, domestically and internationally. From this narrative it is clear that the women's movement was crucial in expanding the participation of women in the United States armed forces by successfully presenting women's struggles, demands, and professional aspirations within two larger important policy issues: equal opportunity and manpower shortage. Joshua Goldstein roughly confirmed this argument when he wrote that:

> the integration of women in the U.S. military came about from a combination of changing norms about women's roles, in turn growing out of the women's movement, and the practicalities of a large, post-draft military in need of personnel. Once women began to integrate in larger numbers after the Vietnam War, practical experience showed this to be a cost-effective way to maintain desired force levels and showed resulting problems to be manageable.[149]

Providing greater equal opportunity in the workplace to women was meant to mobilize the American public, rally their support and start changing norms regarding women's roles in labor in general. Because the issue was articulated within this larger context, today American gender integration in the military is slowly becoming a reflection of gender integration in the civilian labor market. The second policy issue, on providing military services with competent and skilled recruits at all times to prevent manpower shortage, was meant to appeal to the national security community, and in particular the Department of Defense and the White House. Situating gender integration within a broader manpower debate forced policymakers to rethink the military profession as a largely man's realm. While the Vietnam War damaged the image of military masculinity, the Reagan years repaired it and limited women's movement's maneuvering space. However, the Clinton administration's revival of gender integration in military debates, coupled with the end of the Cold War and new military operations framed in terms of compassion and humanitarianism, brought the biggest policy change. Unlike in the United States, political parties and legislative context were a lot more important in the way gender and equality were conceptually framed in Italy. Discussion about gender equality in workplace policies and the elimination of discrimination in Italian society continue to be marred by controversy and political scandals. Institutional reforms were initiated within the military, in those first barracks that allowed a gender integration experiment to take place without much assistance of any other political movements, feminist or not. Most found the issue objectionable and even repulsive. Without aspiring female soldiers fighting the legislative and public opinion battle, the Italian armed forces would not have been the same. However, Italian women's groups are not political parties and did not put the issue on the legislative agenda without allies. As the old left crumbled, and new left political parties became more receptive to NATO's and the European Union's understanding of both military modernization and gender equality, the new women's groups were able to articulate their demands. The opposition by the remnants of the old left and radical feminist remains, but their political

---

149   Joshua Goldstein, email message to author July 27, 2009.

influence has withered. One can safely conclude that in both the United States and Italy, conversations regarding gender integration in the military have always involved a significant domestic political debate that involved military women's associations seeking a bigger inclusion. While that debate in the United States involved women's associations outside the government, in Italy such associations have delayed gender integration policies. Yet, it is important to note that although in both cases the debate opened up as a question of gender equality, these groups found supporters among political and military brass seeking to replace soldiers due to abolishment of conscription and modernization. It was within this changing military environment, in the 1970s and 1990s in the U.S., and late 1990s in Italy that gender integration policies were finally enacted. States need soldiers to fill the ranks of an ever more sophisticated and technologically advanced service, not just men with masculine strength. While the road toward genuine gender integration remains long in both states, the notions of military masculinity are seriously jeopardized by the successes of women.

Chapter 5

Gender Integration in the New Members of NATO: Case Studies of Poland and Hungary

Introduction

While much has been written in the field about women fighting alongside men in both Western Europe and North America, there is little comparable literature in Eastern Europe. However, during both world wars women from this region served as pilots, sharpshooters, nurses and spies, though "collective memory of the world wars is both selective and essentially gendered."[1] In fact, for much of the twentieth century women largely remained out of the history books and out of the ranks of the Warsaw Pact militaries.

From the beginning of this book I have argued that we need to start to study the effects of multilateral military alliances and international security context on domestic decision-making and policy regarding women in the military. In this chapter I illustrate the argument by analyzing the process of integration and expansion of women's participation in the military in Eastern Europe, where international variables have had a much greater impact than in Western Europe and North America, particularly in the years following the fall of the Berlin Wall and the transition to democracy and a free market. As the threat of Communism disappeared in 1989, all of the Western European and North American members of NATO were faced with the new and arduous task of transforming their "ponderous armored formations with hundreds of thousands of soldiers ... to become much smaller and deadlier, light and agile, capable of rapid deployment across intercontinental distances."[2] The same process was to take place in Eastern Europe, though the starting point for these states was very different. They had to first define the role of the military as an institution in their newly democratized societies, and only then pursue professionalization most appropriate to their domestic socioeconomic and political context, as well as the new international security environment in which they found themselves. Policies regarding personnel were passed and implemented quickly and often inefficiently. Therefore, integration of women into the military in Eastern Europe involved different sets of variables and obstacles than in Western Europe and North America.

The argument was confirmed by quantitative analysis which suggested that different factors determine levels of gender inclusiveness as well as the timing of

---

1 Wingfield and Bucur (2006: 10).
2 Adams (2008: 3).

the integration in old and new members of NATO. While in the previous chapter I focused on the old members from Western Europe and North America, in this chapter I study the newcomers in Eastern Europe. It is a narrative account of policy change in Hungary and Poland since the end of the Cold War. These two states were chosen because Hungary has achieved the highest level of integration of women into the military in the region (IGI = 15), while Poland is on the bottom of the gender inclusiveness index among Eastern European states (IGI = 8). This method allows me to compare their different patterns of integration and explore the causes behind them, as I did for the original members.

The two narratives presented here tell the story of the domestic political and economic environment, as well as changing international security within which gender integration in the military policy process begins. What quantitative analysis does not take into account is that Eastern European states were undergoing the process of democratization that brought the constitutional and legislative opening for the conversation regarding gender equality in their military forces to take place. However, I argue that the process was also the result of states' need to conform to NATO's standards regarding equal opportunity before accession. Although these narratives demonstrate that both Hungary and Poland began to open their armed services to women just as they became members of NATO, there is no evidence to suggest that NATO membership alone determines the degree of gender inclusiveness in the military. Rather, it is state's military operational readiness and perceptions of external threats that play a much bigger role after the initial policy.

The political and economic development of Hungary and Poland has been similar since the fall of Communism, but each state has maintained different security policy priorities and different perceptions of external threats. While the relative absence of an external security threat and lack of interest in national security by the government and the people have led to low levels of military operational capability force and high IGI scores in Hungary, the continued threat posed by Russia, geopolitical location, historical experiences and high level of interest in national security has led to high levels of military preparedness and relatively low IGI scores in Poland.

There is no evidence to argue that other variables have had much impact on these IGI scores. While Poland abolished conscription in 2008, Hungary ended it only in 2004, which did not really give its government or the military services that much of a head start when it came to gender integration. Although their IGI scores are so different, both states have negative birth rates. Similarly, the status of women in the military is not a reflection of women's position in the labor market as Poland has a higher percentage of women in the labor market than Hungary, and both have among the highest percentages of women in technical and professional fields of all NATO states, Poland with 61 percent and Hungary 62 percent. Their unemployment rates are also similar—Hungary at 7.83 percent and Poland 7.18 percent.

In terms of domestic political and economic context, contrary to what was expected in the original hypothesis, Poland has a higher percentage of women in the legislature (20.20 percent) than Hungary (11.1 percent). Poland also has a higher percentage of women at ministerial level (26 percent) than Hungary (21 percent). Although Hungary is doing slightly better than Poland in terms of their levels of economic development, both Gross National Income per capita (Purchasing Power Parity) are below the $20,000 mark. In fact, the difference between the two is less than $500. The difference in their IGI scores is also not explained by the activity of domestic women's rights groups as in Western Europe and North America. Such groups range from virtually non-existent to politically weak and irrelevant in both states. When it comes to cultural variables, it becomes somewhat more difficult to compare the states. Poland does have much higher levels of religiosity (over 80 percent) than Hungary (48 percent), but Hungarians seem to have higher numbers of those who believe that men make better politicians (53 percent) than Poles (44 percent). It remains difficult to show that religiosity has had an impact on gender integration in the military, because women's participation in politics or economics does not seem to be affected by it at all as their percentages are relatively high in both sectors. Lastly, Poland is the fourth-highest contributor to UN peacekeeping missions among NATO members, while Hungary ranks fifteenth. Again, this is contrary to the original hypothesis that expected bigger contributors to have greater gender integration in the ranks.

Historical Background

There are only a few historical sources documenting Hungarian women's participation in the military. Most Hungarian schoolchildren will learn of women's heroism in defending the Eger Castle battle of 1552 against Ottoman Empire troops seeking to conquer northern regions and proceed to Carpathia. Once it became clear that the castle might fall, "women joined the battle, gurling down cauldrons of boiling water and tar against the onrushing enemy," and are to this day celebrated as crucial to the victory.[3] They were immortalized in the 1867 *Az Egri Nők* (Women of Eger) painting by Bertalan Szekely that today hangs in the Hungarian National Gallery in Budapest. From the seventeenth century, there is the story of Jelena Zrinski (in Hungarian Ilona Zrínyi), the eldest daughter of a Croatian Ban Peter Zrinski, who married Imre Thököly, the leader of the anti-Habsburg uprising. After his defeat in Vienna in 1683, Thököly retreated to one of his last strongholds, Castle Muncacs, where he was arrested at the end of 1685, and Ilona took over the leadership and continued to fight the Imperial Habsburg Army on her own for the following three years.[4] She and her children are still considered

---

3   Boldizsár (1965: 262–263).
4   Clayton (1879: I, 218–220).

national heroes and patriots who fought valiantly against the Habsburgs and for Hungarian independence.

The first women to serve officially in the Hungarian military joined during the revolution and independence war of 1848–49, where the first female officer served as "order guard," holding the rank of first lieutenant. The majority of women, however, served as nurses and barber-surgeons in the army's medical forces, and there were no laws specifically laying out regulations for women. In fact, the Universal Military Service Law of 1868 only called for every man in the country to serve in the army, with no mention of women's service or duties. During both world wars, civilian women continued to work in administrative or medical positions in the ministry, military headquarters, military hospitals and schools, but not in military units along with men. It was only after the end of the Second World War that there is was a movement toward greater gender integration due to manpower shortage. However, for much of the twentieth century, it was mostly officers' wives who entered administrative, medical and signal fields. Still, they were treated as civilian employees, as they lacked any kind of military training, despite their uniforms and ranks. The primary reason for such integration was the socialist policy that obligated the government to provide employment to officers' wives when their husbands were transferred from one town to another. Such a policy was also a way to ease the struggle of military families existing on a single salary.[5] Throughout the 1980s, the wives typically worked in modest garrisons, usually as part of small military staff units in medical and signal fields. Besides officers' wives, women were allowed to enter the Hungarian military as professional soldiers in four ways: women with college university degrees in medicine, education, law, economics and engineering; women with university degrees in the same fields whose scholarships were awarded by the army; women whose civilian posts in the military were transitioned into officers' posts; and, women who had served as non-commissioned officers became officers after taking the officer's examination, or completing a university degree. It was not until the end of the Cold War that military personnel policies started to really change.

As was the case in all states studied, women soldiers are no novelty in the history of Poland either. Polish women also have a long tradition of military service stretching back to 1830 and a revolt against Russia led by a noblewoman, Emilia Plater. Their participation continued in the national uprisings of the nineteenth century and the struggles for independence during World War I.[6] In 1938, the Sejm, the lower house of the Polish parliament, allowed women to serve in auxiliary detachments, including anti-aircraft, sentry and communications units as well as other services "necessary for defense purposes," and in 1939 the Women's Auxiliary Army Service (Pomocnicza Sluzba Kobiet—PSK) was formed and headed by Maria Wittek. Although largely forgotten today, the actual engagement of women in the national liberation movement was far greater than was stated

---

5   Hungary National Report 2008 to the Committee on Women in the NATO, 1.
6   Goodhart (1920: 157).

in official figures. The commander of the Union of Armed Struggle, Colonel Stefan Rowecki, declared that women in Poland were carrying out the same types of military service as the men, and thus announced that the term Women's Military Service be used already in 1940, but it was not until February 1942 that the Women's Service in Poland was renamed and elevated to the Women's Military Service (Organizacja Przysposobienia Wojskowego Kobiet or OPWK). Only a couple of months later, an order was issued to "make comprehensive use of the women's military service," in preparation for a planned reconstruction of the Polish armed forces. The forces commander sent Elżbieta Zawadzka as his emissary to the Polish government-in-exile to argue for this military service to obtain full legal status. Zawadzka was the only female member of the elite special operations paratroopers Cichociemni. As a result of her trip, in 1943, women were granted the same rights and obligations as males by presidential decree. It was recognized that many underground operations would not have been performed without women, who organized liaison operations, conveyed secrets as messengers and couriers, and delivered illegal newspapers and other printed literature of the Polish Home Army (Polskie Armia Krajowa—AK). A team of female couriers was run by Wanda Kraszewska-Ancerewicznowa, who from 1941 was handed control of the central office of distribution at AK headquarters. In addition, Polish women actively participated in the partisan underground movement, including women's diversion and sabotage detachments called DYSK (*Dywersja i Sabotaż Kobiet*) and women's mine-laying patrols, whose responsibility was blowing up rail lines around Warsaw. Over 60 percent of the couriers during the Warsaw Uprising were women. After it failed, the Germans captured more than 2,000 women soldiers and granted them all prisoner of war status.[7] It has been estimated that almost 5,000 Polish women soldiers perished during the war, which is about 10 percent of those in active service. After the war, many received military decorations. Forty percent of those who received the Silver Cross of Merit with swords were women, as were 50 percent of those who received the Bronze Cross of Merit with swords. Some women even received the Order Virtuti Militari, Poland's highest military decoration for courage, the equivalent of the United States Medal of Honor.[8] After the war, many were arrested by the Security Service, including Wittek, who was imprisoned in 1949 by the Communists, and ended up working the rest of her days in a newspaper kiosk. Similarly, Zawadzka was arrested in 1951 for treason and espionage, sentenced to 10 years in prison, but was released in 1955. Lech Walesa, the president of Poland between 1990 and 1995, appointed Wittek as a brigadier general in 1991, and she became the first Polish woman to attain the rank of general. In 2006, Zawadzka was promoted by President Lech Kaczynski to brigadier general in the army, making her the second and last woman to ever hold this rank in Poland.

---

7   Deck-Partyka (2006: 55–56).
8   Krwawicz 2002).

Communist Legacy: Women's Movements and Gender Equality in Labor

The fact that Zawadzka was given an award some 60 years after the war ended shows that women's war service was not formally recognized by the Communist regime in Poland. Moreover, women soldiers were met with a certain dose of social disapproval which merely reflected a great ambivalence of the regime's attitude regarding gender roles. While women's movements and gender equality in labor informed legislative change in Western Europe, we are left with a question regarding their role in Eastern Europe.

One of the major legacies of the Communist regimes is that women were very active participants in the civilian labor market. Even though Poland always trailed other Eastern European states, including Hungary, female labor force participation was already at 43.4 percent in 1979, much higher than in many Western European states.[9] Similarly, women achieved relatively high levels of professional and technical education. While in 1950, 35 percent of all students enrolled in university were women, by the time the Communist regime was collapsing women accounted for slightly more than 50 percent of all students in Poland.[10] By 1992, the educational level of both women and men was the same, 11.1 years. But unlike in Western Europe, gender equality in the workplace or women's professional and technical training did not affect the integration of women into the military.

The government-sanctioned provision of social welfare benefits and cooptation of civil society during the Communist era left women exhausted and without any real representation. During this period Polish women were expected to find employment and contribute to the "workers' paradise," as well as continue to perform all household chores, raise the children and most importantly wait in endless lines to buy groceries that at times were impossible to find. In terms of associational life, there was only a single women's organization, the Women's League, whose primary mission was to serve the Party's ideology and government. It lacked any kind of political independence that would have allowed it to make demands and participate as a true representative of women's interests.

Despite the recent changes in both Poland and Hungary, women and gender have been routinely overlooked in Eastern European democratic transition. While women participated actively in the social and political movements advocating a return to democracy in both states, democratization often brought disappointment with the way newly elected conservative governments were addressing women and gender roles. For a decade in Eastern Europe, that largely translated into the marginalization of women's interests, and the unwillingness of policymakers to take up gender relations for fear of taking controversial positions.

In Poland, although there are about 200 active women's organizations whose agendas and structures differ greatly, there is a clear division between those that are linked to the Catholic Church and those that are what some might term

---

9   Titkov (1998: 24).
10   U.S. Census Bureau (1995: 4).

more feminist organizations. As was the case in Italy in the 1960s and 1970s, their views differ greatly on women's interests, gender roles, reproductive health and abortion. These divisions have created a diverse, but fragmented, women's movement that has not been able to agree on much. But more than just operating in a divisive associational environment similar to Italy's, some argue that since 1989 Polish women's organizations have been operating in a right-wing political context similar to that of the Reagan-era United States.[11] In the early 1990s, as the door to democratization and a market economy began to open, traditionalists quickly set out to roll back Communist-era gender policies in the new Poland and brought back discussions of gender "appropriate" roles. It was within this process that women returned to military service in 1988, when basic regulations were established to allow them to participate. Although most Polish annual reports to the Committee on Women in the NATO Forces cite 1988 as the year women entered military service, it is necessary to point out that this legislation does not even meet minimal standards of integration compared to the other 23 states in this study. What this legislation did was only allow for a limited women's presence in medical units, where they largely served as physicians, dentists and nurses. They were not allowed to enter military academies or non-commissioned officer schools, nor were they allowed to enlist. The all-male conscription, in place since the end of World War II, remained. Therefore, gender integration as defined and measured in earlier chapters really did not take place in 1988.

During this period the status of women in the society changed dramatically but their position in the military remained unchanged. As one author argues, "the June 1989 parliamentary elections signaled the beginning of a series of threats to women's health and their personal, public and political autonomy."[12] As democracy arrived, even the women of Solidarity were quickly forgotten and pushed into obscurity while men became the new political leaders of Poland.[13] And in 1993, abortion was banned.[14] Many felt that democratization remained unfinished for women, particularly as the Christian organizations (that played a much bigger role in the early days and before the 1993 election) started to push for a more traditional role of women seeking to confine them to the realm of the "church, the kitchen and children."[15]

But before we start jumping to any conclusions regarding the effect that religion or high levels of religiosity might have on women in Poland in general, and more specifically on gender integration in the Polish armed forces, it is necessary to point out that these were simply unrealistic goals of the Church seeking to return women to their traditional roles after more than 40 years of active participation in all spheres of life. A survey conducted in the summer of 1991 shows that

---

11   Fuszara (2005: 1074).
12   Titkov (1998: 27).
13   Penn (2006: 3).
14   Snitow (1993).
15   Titkov (1998: 27).

although 86 percent of Poles thought that the Church was playing an important role in national life, only 7 percent thought it should.[16] In fact, both pro-natalist and pro-family discourse clearly did not resonate with the Polish people as both birth rates and marriage rates have been dramatically dropping since the end of the Communist regime.[17] But the issue of gender equality remains largely unexplored by the society, despite an increasing agreement that it is an important one and that there is a need to address it particularly in terms of lack of it in the public sphere. Yet, gender is rarely part of the legislative debate, just as is the recognition of the importance of equal opportunities for men and women and equal representation by policymakers (Fuszara 2005).

Hungarian transition to the democracy process equally neglected questions of gender equality for many years, as it did not appear on any political party or legislative agenda. Data show that in 1990 the unemployment rate for Hungarian women was at 1.7 percent, and for men 10.3. Within two years, that number of unemployed women was at 8.8, while men's was at 10.8 percent. It should be noted that a decade later, the numbers evened out. While women understood their dire situation, there was still virtually no feminist mobilization within civil society to articulate and demand greater gender equality in society in general, and even, more specifically, in the military (Eberhardt 2003; Arpad and Marinovich 1995). Women were initially too busy surviving. But there is a tendency in Western scholarship to ignore women's groups that are not explicitly feminist, and hence accounts often either mention no groups or single out the Feminist Network as the only organization. Hungarian women's groups have been active since the early 1990s, including the Feminist Network, Women Against Domestic Violence and Equal Opportunity Association and have worked on a variety of issues including abortion rights, maternity benefits and domestic violence. Although Hungarian mobilization was different, as it was often small and connected to and financed by international organizations, such as the European Union through its Phare Democracy Programme, women's movements were present on the Hungarian post-Cold War political conversations (Fabian 2009). The biggest change came in 2002 as the conservative government was replaced by the socialist-liberal coalition, whose agenda finally included gender equality and anti-discrimination legislation. This brought on changes to the Labor Code anti-discrimination legislation, the enactment of Act CXXV of 2003 on Equal Treatment and the Promotion of Equal Opportunities, and a greater focus on gender mainstreaming in all spheres. Along with these changes, there are signs of a broadening of women's groups and their agenda. However, as was the case in Poland, there is no evidence to suggest that there was any involvement by civil society regarding the role that women and men should play in the military. While changes to gender integration policy in the military took place at the time of democratization, and civilian gender equality laws were integrated into military law, that process in Hungary seems to have

---

16   Gudorf (1995: 102).
17   Titkov (1998: 30).

been void of larger societal and legislative debates. The next section demonstrates that structural and institutional changes to the military services in Eastern Europe brought about by the accession into NATO are essential in understanding policies of gender integration in Eastern Europe.

NATO Accession and Military Capabilities

Hungary

Hungary's military history is not one of a grand and victorious military strategy. It was on the wrong side in both world wars, and during the 1956 uprising Hungarian armed forces mistakenly shot at their Soviet colleagues, and in the end completely fell apart as they failed to reestablish order. Until 1989, Hungary was a loyal member of the Warsaw Pact, but as it did not boast a very important geo-strategic location the Pact did not pressure it to spend money on major defense reforms and improvements, as was required of Poland. The lack of interest on the part of the Soviet Union, allied to Hungary's own lack of interest regarding military issues in general, affected defense reforms in years to come, particularly in terms of budget cuts from the late 1980s and throughout the 1990s. By the late 1980s among the Warsaw Pact member states, Hungary had not only the smallest army and air force but also the fewest light armored vehicles and artillery and anti-aircraft weapons. Besides being short on available weapons, Hungarians were lacking in the quality of equipment to hand. According to Western analysts, Hungary's military forces had the lowest combat readiness in the Warsaw Pact and were the least trusted force even by the Soviet leadership in Moscow (Dunay 2005).

As Hungary started to transition to democracy, military elites stayed largely out of the political discourse, as they had for much of the Cold War, because the constitution did not allow soldiers to be members of political parties, and, indeed, prohibited them from participating in politics. The reformed Communists attempted to influence the 1989 elections and further shield the Hungarian military from political forces by separating the Ministry of Defense from the armed forces, which were placed under the command of the president. They hoped the president was going to be their reform leader, Imre Poszgay. The reform Communists lost the elections, but their Defense Reform of 1989 created problems for the next twelve years before the armed forces were finally placed back under the command of the Ministry of Defense (Martinusz 2002). For most of the 1990s, civilians heading the MOD had little or no knowledge of defense, and the minister's post was used to reward political loyalty and to negotiate deals. Most of the ministers not only lacked knowledge but also interest in defense and armed forces. On the other hand, old military professionals also lacked expertise as they had been told what do to by the Soviet High Command for more than 40 years.

The situation throughout the 1990s did not change considerably as defense remained far behind other more pressing issues such as economic restructuring and development and institutional changes to curb widespread corruption and crime. The people of Hungary did not believe that investing in the military was a necessity, and therefore political leaders did not seek to allocate funds to an institution that appeared to be entirely useless and broken beyond repair. In fact, the political parties in power did not even believe NATO would offer them an invitation in the foreseeable future and thus chose to reallocate a big part of the defense budget to improve the Hungarian economy. The government simply felt that there was little need to improve the situation of the armed forces due to the lack of significant threats. As such, gender integration was placed on the bottom of all political agendas, because government members, women's groups and parliamentarians paid little attention to it. As Dunay argues, their disinterest "reflected the largely threat-free geostrategic environment that Hungary finds itself in, as well as Hungarian society's traditional lack of interest in defense issues."[18] Defense spending was cut to the dangerous level of 1.4 percent of GDP in 1995, and 1.2 percent in 1996.[19] The Chief of Defense, the highest-ranking military commander of the Hungarian armed forces, and Chief of Human Services were left with the task of developing a plan that would counter the budget cuts as well as find personnel interested in defense to keep their ranks filled as many professional men were leaving due to better economic opportunities in civilian life. The military leadership was forced to deal with manpower shortage and assign women to positions that were traditionally regarded as appropriate only for men.[20] As the Statute (CX / 1993) on National Defense opened up certain officer, non-commissioned officer and regular positions to volunteer soldiers in order to strengthen the professional character of the Hungarian Defense Forces, gender integration was to take place under the command of the Chief of Defense, or the military leadership, and not under the civilian leadership of the Ministry of Defense and parliament.

Although domestically security and military policy were on the bottom of the political agenda, the Hungarian government was actively seeking membership in one of the most exclusive clubs—NATO. In the Hungarian case, decisions to start phasing out conscription and to begin gender integration coincide with the establishment of NATO's Partnership for Peace (PfP) program in 1994. This program, aimed at strengthening security ties between NATO and Eastern European states, helps educate and prepare applicants' armed forces for their participation in the alliance. That same year, women were finally allowed into military academies, though they were initially limited to studying logistics, finance, signals, radio-reconnaissance, and informatics. In 1996, all positions, including combat assignments, were opened to women and their involvement in the military expanded to include anti-tank destroyers and airborne infantry posts

---

18   Dunay (2005: 23).
19   Sherr (2004: 69).
20   Bolgar (1999: 1).

in 1997, and all faculties and training programs in 2005.[21] By 1998, the first 29 female officers completed their education at the Bolyai Janos Military Technical College in signal, finance and radio-electronic reconnaissance. In 1999 the Hungarian Armed Forces Equal Opportunities Committee, representing middle management in each service, was created in order to report information and ideas to the Chief of Defense regarding gender integration.[22]

In 1999, Hungary became an official NATO member, and it was clear that the government and the people of Hungary would have to start paying attention to defense issues. But the government only realized how dire the situation was after it took the armed forces six weeks to identify and assemble 300 qualified soldiers that were to become part of NATO's Kosovo Force.[23] The crisis of manpower became a source of embarrassment and, although the Ministry of Defense civilian leadership recognized it as such for more than 10 years, it was responsible for withholding finance necessary for basic functioning military capabilities. The ministry failed to deliver the funds because the Hungarian people did not support the investment and never really saw their armed services as an instrument of foreign policy or even national security.[24]

Another problem affecting the personnel of the Hungarian armed forces was the fact that a large portion of civilians in the defense sector were and continue to be political appointees, which does not allow for much stability as they are demoted and dismissed depending on election outcomes. Besides the lack of permanency, the political appointees often lacked any knowledge of the defense sector or genuine concern with military personnel issues as the positions were only serving as springboards for other, more important political positions.[25]

By 2001, when defense reform finally made the highest military commander, the Chief of Defense, subordinate to the Ministry of Defense, only 52 percent of total authorized enlisted slots were actually filled and this crisis within the services was actually providing an opening to women.[26] As a part of the new reform, in November 2003, the Committee on Women of the Hungarian Defense Forces (HDF) was established by the joint decree of the Administrative State Secretary and the Chief of General Staff of the Ministry of Defense of Hungary. This committee's aim is to analyze women's position in professional and contracted service, study problems regarding female personnel and propose solutions to the Ministry of Defense.[27]

---

21   Hungary National Report (2008: 2). Available at: http://www.nato.int/ims/2008/win/reports/hungary-2008.pdf (accessed December 2009).
22   NATO (1999a: 32).
23   Sherr (2004: 74).
24   Martinusz (2002: 286).
25   Sherr (2004: 79).
26   Martinusz (2002: 292).
27   Hungary National Report (2004: 1).

A year later, the HDF became a voluntary, professional army consisting of professional and enlisted soldiers, and their role became to defend the sovereignty and territorial integrity of the Republic of Hungary and to contribute to the collective defense of NATO. Recruitment problems remain, but the numbers of women continue to rise, as does their proportion at all levels. According to Andrea Szabo Szabóné, a lieutenant colonel in the Hungarian military, "The military, the defense forces have no gender. Military only has power, the capability to fight, authority and social prestige. The unity of the defense forces does not know gender or age. There are female soldiers and there are male soldiers. Professional competence, capabilities and qualified knowledge are what matter."[28] Similarly, in March 2011, at the Scientific Conference of the National Defense University celebrating 20 years of the HDF, Dr. István Simicskó, the Ministry of Defense Parliamentary Secretary, recognized that more and more women practiced this very serious and "manly profession ... and stand on their own." He added that the Hungarian military understands that physical force is no longer enough, so he suggested that Hungary expects "the people who are mostly mothers and wives, and also enrich us with their skills at their profession."[29]

In fact, this thinking has informed much of the gender integration policy in Hungary, as there were no heated and controversial debates marring the process—only an understanding that without gender integration there is no military modernization now required by NATO. Women make up 21 percent of all junior officers, and every second OF-2 officer (captain) and every third OF-3 (major) officer is a woman. The percentage of female non-commissioned officers from OR-6 to OR-9 (several ranks of sergeants) is between 22 and 42 percent, which is the highest among all NATO members.[30] The main complaint is that the majority of women remain working in medical, administrative and human resource fields, but their numbers are slowly starting to rise in other areas. For example, women now make up over 34 percent of air combat support and 11 percent of artillery, but there are no women in the Special Forces. Women are present on almost every level and in every department of the Hungarian military, including in overseas operations.

As part of modernization, the HDF were dramatically reduced in size, including some of the units where women traditionally have been overrepresented, such as the medical service, financial, and logistic units. However, this change did not negatively affect gender integration, because the majority of personnel who were downsized or retired were male employees.[31] All new training programs for women are organized the same way as those for male soldiers. In the military faculty at the National University of Public Service and NCO Academy, all courses are open to both women and men, without discrimination in entrance

---

28   Szálkai 2013.

29   "Katona-nő és a haderő Tudományos konferencia a ZMNE-n". March 24, 2011. Available at: http://www.honvedelem.hu/cikk/25008 (accessed June 2012).

30   Hungary National Report (2008: 4).

31   Hungary National Report (2008: 3).

or studies. The Hungarian government has also developed initiatives aimed to make their military more female-friendly. The Strategic Plan for Gender Equality (2010–20) is implemented by means of three-year action plans in each relevant domain, including the armed forces and it is meant to promote gender equality and transforming the traditional approach to gender relations. The National Action Plan presents specific actions to allow for opening up traditionally male-dominated areas to more female soldiers and prevention of sexual harassment by creating a system of legal remedy. In addition, mainstreaming human rights and gender into crisis management activities is being developed by Hungary, with the participation of the Ministry of Human Resources, Ministry of Defense and Ministry of Justice and Law Enforcement. Hungary also allows for the women's section of the Military Trade Union that represents servicewomen and fights for rights and allowances of (military and civilian) women within the armed forces. Although there are no gender advisers in the HDF, gender is incorporated in pre-deployment training and there is an appointed person in each detached unit whose responsibility is to promote equal opportunity and rights. Family policies have also been addressed to allow single fathers and mothers to not be obligated to complete 24-hour duty until a child is 6 years old. The same applies to military couples if one of the spouses is on a mission abroad. Moreover, the military offers economic safety to women, and women are interested in joining the military because unlike the private sector, the armed forces offer a stable income and employment that can be safely maintained during maternity leave.

Hungarian military reforms have taken place during a difficult process of democratization and accession to NATO and the EU. Few security concerns and decades of general disinterest in defense by the people and the government have left the new HDF with low operational capabilities and without the ability to operate and sustain missions abroad. With the constitution and democratic principles providing for gender equality in all spheres of life, and NATO's demands regarding modernization, gender integration in the military was a rather painless and necessary part of the military personnel reform process. It has allowed for a dramatic expansion of their women's role in the services in a very short period. Hungary still faces numerous challenges to its military modernization program, as reflected in the 2008 Hungarian defense budget, set at 1.17 percent of GDP, which continues to be well below the NATO target of 2 percent. Despite financial problems and political transition at home, Hungary has been firmly committed to NATO from its accession, and has actively participated in a number of different operations, including the war in Afghanistan, where two of Hungary's female soldiers have lost their lives. While no gender integration in the military is an easy process, Hungarian policymakers and the state's military have made enormous strides, and set an example for the rest of the alliance.

Poland

Throughout much of the 1990s, there was no discussion regarding the expansion of women's numbers or roles in the military ranks as other more pressing political and economic problems associated with the transition were on top of policymakers' agendas. However, unlike in Hungary, political and economic problems never completely overrode the issues regarding national security. On the contrary, from the beginning, Poles were keen to improve their armed forces and actively sought membership in NATO. By the mid-1990s, the Polish government sought to show its commitment to the alliance by joining military exercises with NATO troops, including a Polish assault battalion's participation in the NATO-led peace implementation force in Bosnia.[32] At the same time, Hungary was unable to help and even in 1999 was still scrambling to find a meager 300 troops to join NATO operations in Kosovo.

In 1996, Polish President Aleksander Kwasniewski reaffirmed "Poland's continuing and unequivocal aspiration to become a full-fledged member of the North Atlantic Alliance at the earliest possible date" and sought to speed up the accession by pointing out that Poland is working hard to "achieve full military interoperability with the Allied forces, commencing with command, control, communications, and intelligence as well as management of air-defense systems."[33] But why was Poland so much more committed to becoming a full member of NATO than Hungary? And did this have any effect on gender integration in the military? According to Polish Foreign Minister Dariusz Rosati, becoming a member of the alliance was the most important issue on the government's agenda in the mid-1990s because:

> it is also clear that Poland's political and cultural development, its economic prosperity, and, ultimately, its military security depend on rebuilding and cementing its ties with the West—the cradle of our culture and statehood. Poland will feel fully secure only as an integral and indispensable element of the European family of nations, whose cultural roots, values, and aspirations we share. This is the main reason why political, declarative guarantees of security are of no interest to us-they simply fail to provide what we are looking for. For us, NATO enlargement means much more than extending security guarantees to new nations. In Poland's view, enlargement is the only realistic way to build a new, effective security architecture for Europe and to overcome the divisions of the continent.[34]

Poland and Hungary differed in their approach to the alliance: as Poland sought to modernize and reorganize its massive military, Hungary was caught up in political

---

32   Alexander Kwasniewski, President of the Republic of Poland (1996).

33   Kwasniewski (1996).

34   Rosati (1996).

and institutional chaos. Still, both were invited to join NATO at the same time, in March 1999. What followed was a series of changes in policy regarding women in the military at the urging of the Committee on Women in the NATO Forces.

The same year, Poland established the first Council on Women in the Polish Armed Forces, and sent delegates to the committee to participate for the first time. In fact, 1999 was a year of remarkable change as women were now able to attend military schools and service academies, and were integrated in the services under the command of the Chief of Defense. They were to work and train with their male counterparts and to become subject to the same chains of command, standards of performance, and discipline. Also in 1999, regulations were passed to allow women to become either permanent regular soldiers or voluntary contract soldiers. Only women with a university degree or with a high school degree could apply. Those with a university degree were to become part of the officer corps, while those with secondary education were allowed to join the warrant officer corps or non-commissioned officer corps.[35] According to Commodore Bozena Szubińska, the chairwoman of the Council for Women's Affairs in the Polish Army, the academic year 1999/2000 provided a fundamental breakthrough "and was a consequence of our accession to NATO.[36] External pressures and international standards have proved to be more effective to work for equality between women and men than the mechanisms provided for in Poland to defend the rule of law and compliance with the Constitution and laws of lower order." She argues, however, that the change took place in an environment of continuing democratization, in which new sources of independent media contributed to raising public awareness and strengthening women's activity. This in turn allowed for a growing expression of feminist movement which began to demand not only the relevant legal provisions in the sphere of equality, but also concrete actions in real life. However, in the Polish context, she identifies strong external pressures that have committed the Polish government to meet the appropriate gender norms and standards, respect for human rights and equal treatment for the elimination of all forms of discrimination as they are foundations of the developed democracies. Hence, while on the one hand domestic social and political transformations have provided a more conducive environment, on the other, "joining NATO, where the requirement of democracy is the equal participation of women in all areas of human activity, including in the security sphere," proved to be central to gender integration in the military.

Yet, while integration took place quickly to fulfill the obligations and responsibilities under the NATO charter regarding modernization and equal opportunity in the ranks, Poland has not followed the Hungarian path of integration. In fact, Szubińska recognizes that the pace of gender integration in the military has been much slower in Poland than in other countries of the alliance. She notes that the transformation of the political system, accession to NATO and the consequent reconstruction and modernization of the military have

---

35   NATO (1999b: 44).
36   Personal correspondence with author, September 9, 2010.

reduced a nearly 500,000-strong army first to 180,000, then 150,000, to current levels of around 100,000. Adaptation to new threats, challenges and needs of the international security environment have forced Poland to phase out the all-male conscription. The abolition of conscription and positive trends in the modernization of forces, Szubińska argues, did not make the Polish military more conducive to the development of gender integration.

The reasons for it might be in the military's continued and open declaration of preference for male recruits. The 2008 report to the Committee on Women in the NATO Forces states that:

> as women in Poland are not subject to conscription, their calling takes place in case of "justified need of the Armed Forces" if they possess particular qualifications or skills. Hitherto recruitment to professional Private Corps aims at drafting of over-term soldiers and soldiers who served in the military thus men have easier access to this formation.[37]

By the end of 2011 (the Department of Defense Personnel), 1954 professional military women have served in the Polish Army, which represents 2.11 percent of all-professional soldiers. They are privates, NCOs and officers and are represented in the land forces (737), air force (391), navy (190), Special Forces, military police, the Inspectorate for Armed Forces Support, as well as in the structures subordinate to the Inspectorate of Military Health Service. Numbers are increasing and in the last year, an additional 203 women joined. In 2011, 140 women participated in peacekeeping operations, representing approximately 2.6 percent of participants in all missions.

In 2012, the Polish Army will start its promotional activities aimed at increasing women's military participation from 2.11 percent to 5 percent. Until recently, their recruitment depended on the number of candidates in military academies and schools as announced by the Minister of National Defense every year, but today there are no such restrictions. Yet, despite the improvements, the Polish Army does not have gender advisers, and "gender" is part of the operational planning process. The occupational limitations continue to exist and are quite different from those imposed in other NATO states. Women are not to serve in posts that require physical effort and transport of heavy goods and forced body position; in cold, hot or unsettled microclimates; in posts that expose them to noise, vibration, electromagnetic fields, ionizing and ultraviolet radiation and monitor screens; underground and at high altitudes; posts that expose them to high or low pressure, hazardous biological elements and chemical substances; and posts that could cause them grave physical or mental injury.[38] No other NATO state is as specific as Poland when it comes to defining posts that could be strenuous and detrimental to women's health.

---

37 Women Military Service in Armed Forces of the Republic of Poland (2008: 4).
38 Women Military Service in Armed Forces of the Republic of Poland (2008: 6–7).

When it comes to sexual harassment, Poland has also been lagging behind. "Most often the victims of sexual harassment are the students of military academies. They are afraid of losing the chance to serve in the army, so they do not notify the prosecutor about the offense," says Szubińska. But what the government is doing seems little and inadequate to fight sexual crimes. The Ministry of National Defense financed the publication "Co-educational Army. Gender Equality as an Issue for the Teaching Staff in Military Schools," while in 2007, two educational films—one on discrimination and fighting societal stereotypes of women's role and the second on sexual harassment in the military—were produced.[39]

The data of the Military Courts Department in the Ministry of Justice show that in 2003, only seven soldiers were sentenced for sex crimes, in 2004 none, and in 2005 eight. According to Polish criminal law, rape with a special cruelty is punishable by up to 15 years in prison, "ordinary" rape up to 12 years, and harassment up to 3 years imprisonment. Yet military courts seem to be quite liberal and soldiers receive a suspended one- or two-year prison sentence.[40]

Today the Polish armed forces number about 140,572 troops divided among an army of 87,877, an air and Defense Force of 31,147, and a navy of 21,548. All Polish male citizens were subject to a 12-month term of military service before the period was reduced to nine months. Only in 2008 did Poland start to move away from conscription and currently is in the process of transition to an all-volunteer force. Although financially Poland is not always able to support the restructuring and modernization of its military, it has been able to move forward with U.S. assistance on acquiring 48 F-16 multi-role fighters (that were delivered between 2006 and 2008), C-130 cargo planes, and High Mobility Multipurpose Wheeled Vehicles (HMMWVs or more popularly known as Humvees).[41] It continues to be the strongest military in Eastern Europe and with the highest operational capabilities. Unlike any other state in the region, Poland took over command of a sector in Iraq in September 2003 with 2,000 deployed soldiers. In addition, 7.5 percent of the entire Polish Army is available for deployment on operations.[42] In fact, Poland is represented in a second tier of conflict capabilities along with the other Western European states: Netherlands, Italy and Spain.[43] No Eastern European state is represented on this level. Although it has been 10 years since NATO's expansion, Poland still sees the alliance as a shield against Russia. "Russia was, is and will remain unpredictable. It won't stop being a problem. We saw that with the Georgia crisis. Putin has clear ambitions to rebuild a military superpower," said Polish security policy expert Jan Czaja.[44] Today, Russia is

---

39   Women Military Service in Armed Forces of the Republic of Poland (2008: 7).
40   Agnieszka Mrozik (2006).
41   U.S. Department of State (2009).
42   Lindley-French and Algieri (2004: 34).
43   Lindley-French and Algieri (2004: 31).
44   "Ten years since NATO expansion, Poland sees alliance as bulwark but worried about Obama." *Associated Foreign Press*, Warsaw, Poland, March 19, 2009. Available

in resurgence and Poland is concerned about its security once again, especially after violent actions in the Caucasus, which came as a great justification of traditional Polish distrust of Russians. In the aftermath of the Georgian conflict, the United States and Poland signed an agreement to place an American missile defense base on Polish soil. Polish government officials supported this agreement wholeheartedly because they felt this would strengthen the commitment of the United States to defend Poland against Russia. "Poland and the Poles do not want to be in alliances in which assistance comes at some point later—it is no good when assistance comes to dead people," said the Polish prime minister, Donald Tusk, on Polish television.[45]

Since Poland entered NATO and was required to join the Committee on Women in the NATO Forces in 1999, virtually no additional policies expanding gender integration have been passed by Polish legislators. With Russia flexing its muscles since President Vladimir Putin's arrival in 2000, and strong and unequivocal support of Operation Iraqi Freedom, Poland has been forced to tailor its military personnel needs accordingly. Defense spending accounted for 1.95 percent of Polish GDP in 2008, making it one of the biggest military spenders relative to its size not only in its neighborhood, but in the whole of Europe. Hence, we can see that although NATO membership has forced Poland to open the door a crack to women in the military, its conscription and security concerns have slowed down gender integration. Yet there is hope that the Polish military will soon be forced to change its policies. As the last conscripts completed their service in September 2009, and with close to 38,000 conscripted soldiers ending their military participation, up to 20,000 additional professional soldiers will be needed to make up the difference. The biggest problem for this new professional military will be to fill its ranks by attracting volunteers. Wladyslaw Stasiak, head of the National Security Bureau, which gave military advice to President Lech Kaczynski, argues that this speeding up of reform "coincides with a lack of a thorough action plan, solid calculation, indication of sources of financing, as well as a motivation system for volunteers willing to put on uniforms."[46] Extending the invitation to both women and men might just be an answer to their manpower problems.

Conclusion

Discussion regarding gender integration in the Eastern European military services took place under different political, economic and international conditions than those in Western Europe and North America. The democratization process did not bring about the political opening to women's groups, and, in fact, questions

---

at: http://www.russodaily.com/reports/Ten_years_since_NATO_expansion_Poland_sees_
alliance_as_bulwark_999.html (accessed November 2009).
   45   Shanker and Kulish (2008).
   46   Nicholas (2008: A5).

of gender were largely left untouched by the early conservative governments. In the Eastern European context, gender integration in the military was not part of some larger discourse on gender norms, but rather was passed in connection to the NATO accession, budget cuts and military force and operational capabilities adjustments associated with that accession.

For the most part, the Hungarian government and the public are largely uninterested in investing in their military services or creating a sophisticated force able to participate in major military operations. Although lack of reform has left its forces in shambles for most of the 1990s, once Hungary became a member of NATO, it became clear to policymakers just how empty their ranks were, and they quickly opened them to all who were qualified to serve—including women. Today, Hungary is seeking to modernize its new all-professional force, yet it continues to be on the bottom of the military-readiness scale. While domestic legislative changes, including the Labor Code and the CXXV Act, have further reinforced Hungarian commitment to gender equality, we can expect integration to keep expanding.

Poland, on the other hand, haunted by its bitter history of Russian and German conquest, has been more acutely aware than most European states and NATO members about questions of national defense. Defending one's land is an integral part of the Polish narrative, yet it's almost absent from Hungary's. Polish policymakers have had to tailor both their national security policy and their military manpower needs according to their perceptions of a new and more aggressive Russia. The age-old enemy has caused both the people and the government to seek greater cooperation with NATO, but also to keep and heavily invest in their old mass military model, including all available males and virtually no females. While NATO has exerted just enough pressure to open the Polish military to women, gender integration has not been accomplished. It seems that the legislation allowing more women into the ranks was more of a down payment for Polish membership in NATO, and once accomplished, discussions of gender integration were overshadowed by a lack of modernization and the Polish mass military model. Centuries of rivalry and invasions have left Poland suspicious of Russian intentions, while post-Cold War politics have left Poland tied to the EU and NATO military policy. The end of conscription is the end of an era for the Polish state and its military, but it might just finally be a chance for the gender integration process to really begin.

Chapter 6
Conclusion

Every nation in the world has a story of a brave woman who fought and died for her people and her land. From Mohammad's wives to Joan of Arc to Molly Pitcher, women and men were united on the battlefield, but most of the time as a war ended, men's heroism and accomplishment became historical accounts, and women's often barely earned a footnote. States have for centuries produced, reproduced and relied on military masculinity as an institutionalized form of exclusion of all other masculinities and femininities. As such, the warrior was on the top of all societal gender hierarchies. Military masculinity subordinates all other cultural and societal gender constructs and requires femininities to exist in order to produce this highly gendered organization.

Today, women's stories of gallantry and sacrifice have finally become headline makers, particularly since the end of the Cold War and especially during the wars in Iraq and Afghanistan. The dramatic changes in the international security environment have prompted reorganization of military forces and renewed questions about the proper role of the military, the role that state plays in shaping the military profession and about state's ability to balance military manpower demands within the context of democratic society. All NATO member states were faced with the same new international security environment, some better, some less equipped to deal with the new and changing realities of conflict, including insurgencies, ethnic warfare, and terrorism. Military and civilian leaderships were forced to find answers and review operational and strategic capabilities according to perceived threats and NATO requirements. This process involved a substantial military reform and transformation of all NATO member states, particularly in regards to modernization and force structure. The majority of the states decided to leave the mass military model in favor of an all-volunteer force, staffed with professionals with technical skills. Adjustments were often followed by heated legislative debate over defense budgets and who these new soldiers were going to be. As the skill difference between military and civilian labor sectors started to narrow, all states understood the need to broaden recruitment to include both women and men. The military profession could no longer stay isolated and separate from the rest of the democratic society it claims to protect. Initially restricted to trades that at the time were considered feminine, women's role in the military has dramatically expanded in the past couple of decades to include combat, the ultimate expression of masculinity. Gendered military ideological perceptions affect the process of policy change, from the difference in physical training requirements between men and women to educational programs to the types of roles and positions open to women. Yet while some states have made gender integration a central pillar of their military personnel policies, others have continued to maintain gender restrictions. This book has sought to answer why there are such differences in the degree to which gender integration policies have been enacted, and as such is the first in political science to ask that question in a comparative and systematic way.

While security questions have been traditionally addressed by the international relations scholarship, the analytical focus has primarily been on causes of war, state's grand strategies, military operations and technology. The focus on systemic variables and lack of domestic policy analysis has left the analysis of gender integration in the military untouched by the scholarship. Gender as a variable was deemed irrelevant until the critical broadening of the security agenda and the emergence of feminist security studies brought it into international relations as an analytical lens. In the past two decades the feminist scholarship has questioned and exposed masculine conceptualization of security devoid of women's experiences, and representations of leaders and militaries as the embodiments of hegemonic masculinities. They demonstrated that traditionally, both military and our studies of it, have largely constructed the profession in terms of its ties to masculine attributes such as physical strength, aggression and heterosexuality. In becoming a soldier, one was renouncing all feminine attributes, as masculinity informed men how they ought to behave in the military. Men were trained to be masculine soldiers, loyal to their nations and their fellow brothers in arms.

The construction of the military, by the feminist security scholarship, as an inherently masculine institution has allowed for little analysis of the factors that prompt state's policies of gender integration as they were interpreted as another form of patriarchal militarization. Such structural explanations did not study the effect of domestic political actors, particularly women's movements as agents of change; while international gender mainstreaming efforts have often been portrayed as legitimizing gender stereotypes. On the other hand, liberal and difference feminists' focus on domestic legislative process has not been systematically comparative and has left international variables unexamined.

In 1995, Mady Segal was the first sociologist to study causal factors concerning the integration of women into the military and expansion of their roles. Her seminal work was followed by a handful of other scholars from the same field that expanded the list of possible causes, but they largely followed Segal's conceptualization and method. While their work has shaped my approach to the question of gender integration, my conceptualization and arguments differ from theirs as I draw on both the literature in international relations, in particular feminist security literature, and comparative gender policymaking literature. The explanatory model is based on previously proposed models and conceptual categories developed by military sociologists. From the beginning of the book, I have tried to make this a scientific project, with clear and discernible hypothesis, empirical testing, and evidence-based conclusions—so that we can claim "progress" in our understanding of why some states integrate gender and why others do not. This has allowed me to reconceptualize key terms to situate both the discussion within theoretical literature of security and gender, and add a necessary deductive rigor to offer a simpler, concise and parsimonious explanation of gender integration in the military.

By studying 24 NATO member states' policies of gender inclusiveness, and by combining both large-N quantitative analysis and process tracing case studies of the United States, Italy, Hungary and Poland, this book sought to build the theoretical model that explains gender integration. First, I empirically tested four variable categories proposed by the previous models: military manpower, domestic political and economic factors, cultural factors, and international security context. After testing an exhaustive list of structural, institutional, cultural and systemic variables, I have found that civilian policymakers and military leadership no longer surrender to parochial gendered division of the roles, but rather integrate women (1) to meet recruitment numbers due to military modernization and professionalization, (2) to meet demands of domestic women's movements and (3) to meet state's responsibilities under international agreements regarding gender equality and gender mainstreaming in the military.

The quantitative analysis revealed, however, that such generalizations can be somewhat deceiving. Despite their shared histories, similar political and economic arrangements, and membership in NATO, the analysis showed that there are differences in patters on gender integration policymaking in Western European and North American (old member states) on the one hand, and Eastern European (new member states) on the other. It demonstrated that while there are no significant differences in the degree of gender inclusiveness between the two regions, the timing of integration in Eastern Europe is directly correlated to the timing of states' accession into NATO. After repeating the test while controlling for region, the analysis demonstrated that there are different factors that help explain the extent of gender integration in each region.

Case studies allowed me to trace the gender integration in the military policy process in old and new members of NATO, and to examine in greater detail results what the quantitative analysis revealed. In the United States, active women's movements directly lobbying and working with Congressional members regardless of ideological differences succeeded in opening and extending gender equality in the U.S. military. By presenting their demands within a larger context of equal opportunity in the work place and equal citizenship they allowed for qualified women from the civilian sector to "spill over" into the military sector. This case confirmed that in the original NATO member states in Western Europe and North America, independent movements, or movements that are not functioning as a party wing or political caucus but representing women's choices, were crucial in articulating demands and mobilizing resources to pass the legislation.

Although the women's movement in Italy was slow to start organizing around the issue, similar to their American counterparts, they organized, lobbied and formed strategic partnerships with legislators who would present the bill at a time when the rest of the government members could not afford not to support it. In the U.S., it was the time of the Equal Opportunity Act and abolition of the draft, and in Italy, it was the time of reorganization of the party system, the rise of the new left and new right, as well as the abolition of conscription. These events opened a window of opportunity for women's groups to pressure the government to change the legislation regarding their full access to military services. Therefore, the analysis of Western European and North American states demonstrates causal mechanism at work by tracing the process of the policy change that has developed through domestic institutions, and the extent that they provide arenas for feminist actors inside and outside government to mobilize.

On the other hand, the story of women in the Hungarian and Polish militaries has not been determined by domestic women's movements, nor did the abundance of women in professional and technical fields generate the same spillover effect. As there was no interest in having the state address gender-related issues in the military, and as there were no large women's groups demanding equality, nothing happened until the government was obliged by NATO. Demands to open up militaries came from NATO as part of their standardization, professionalization and modernization efforts. Therefore, in Eastern Europe, it was demands from the international military alliance that forced the domestic policymakers to include gender integration policies as part of their accession process. However, that does not mean all states adopted all of NATO's gender mainstreaming measures. What the narratives showed is that the degree to which women were integrated in Eastern Europe is largely conditioned by the perceived levels of threat by the nation, and the corresponding readiness and operational capabilities of state's armed services. The threat of potential attack by Russia has kept the Polish military heavily armed, conscripted, and with a rather low level of gender inclusiveness. On the other hand, a relatively safe and calm international security environment has helped turn the Hungarian Defense Forces into one of the most gender-friendly employers.

There is another important detail exposed by the case studies and that is that states' policies on gender integration were long overdue in all states. Military masculinity persists and it continues to present itself as an antiquated but very dangerous obstacle to further integration. States have been forced to react, often after they find themselves accused of holding onto "men's" militaries that are no longer relevant in an ever-changing security environment. Yet the vital element of operational and strategic leadership is the ability to successfully organize state's military services. The inability to adapt renders the strategy and the force weak.

And in a world where wars are requiring technological mastery, cyber skills and knowledge of foreign cultures and language, those who maintain low levels of gender integration in the military will have a difficult time protecting their nations. The process of modernization has required states and the public to rethink the analysis of the binary construction of male/female and the corresponding appropriate masculine and feminine behaviors in the military. It is a complex and bumpy process as states have for centuries relied on a military masculinity to justify their wars, domination, violence and militarism, although, we might have to endure "culture wars" by the pundits and social commentators whose debates are often based on faulty premises. The fact is, women and men fight alongside each other during war, complete the same difficult tasks and are certainly chipping away the notion of traditional military masculinity.

Appendix A

Data Collection and Measurements

Measuring Gender Inclusiveness: All of the data here are collected through direct correspondence with a country delegate to the NATO Committee on Gender Perspectives, and through review of reports published on the Committee's website.

Manpower Need

Data Collection: Primary data regarding personnel and accession policies in the NATO member states were collected from NATO Committee on Gender Perspectives yearly reports. These are available for the years 2000–2012 and include information regarding the evolution of the forces and the specific policies regarding integration and expansion of the role of women in the individual states. It is important to point out that not all states have contributed the same quantity or quality of data, and therefore at times the data from the aforementioned reports were combined with outside sources. These include official Ministry of Defense websites, the Rand Corporation and International Institute for Strategic Studies *Military Balance* publications that provide additional information such as statistical data regarding armed forces in individual states, final texts of legislation, directives and decrees on draft, professionalization and modernization of the armed forces and women's services.

Birth rates for all case studies were obtained from the United Nations Development Program (UNDP) country reports. In addition, analysis of the female labor market and female employment rates in technical and professional fields was conducted by utilizing the data provided by the UNDP in its Human Development Index (HDI). Because of the nature of this study and the length of this project, I have opted to use economic data prior to the beginning of the global recession in 2008. After spending countless hours seeking to find the most up-to-date information, I realized that data were changing so rapidly and dramatically, but policies on women in the military did not see too much change. Women in the military were not a particularly interesting subject, given that within a very short period of time in Spain and Greece over a quarter of population were abruptly made redundant, and in most other states numbers at least doubled. The United States Bureau of Labor Statistics and United Nations Economic Commission for Europe (compiled from national and international official sources such as EUROSTAT, OECD and CIS) have provided unemployment rates in all countries for 2008. The information in this study was also supplemented by research published in academic articles, reports and books analyzing military policy in individual states, particularly in the qualitative chapters.

Domestic Political and Economic Context

Data Collection: For all states, the statistics regarding the numbers of women in lower and upper levels of the legislative were collected from the 2008 Inter-Parliamentary Union report. The UNPD Human Development Index provides statistics regarding the percentage of women in ministerial positions, and states' level of development, as measured by the Gross National Income per capita (Purchasing Power Parity) in 2008, was provided by the World Bank. To measure women's participation in the labor force I have used the data from both the OECD and World Bank. The OECD provides 2008 data on the women's labor force as a percentage of total population aged 15–64 for its member states, while the World Bank has provided me with 2007 data on Bulgaria, Lithuania, Latvia, Romania and Slovenia.

The existence of autonomous women's movements was studied only among the original membership of NATO states, because these are the countries that have experienced both second and third waves of feminist movements. Studying, comparing and coding women's movements is not the easiest task, but this one had already been accomplished by Laurel Weldon for her project on government policies on violence against women. According to Weldon's criteria, autonomous women's movements "must be self-governing, and must recognize no superior authority, nor be subject to the governance of other political agencies" (2002: 79). In qualitative chapters, this book studies women's institutional and associational groups that were involved in the framing of the legislation in the United States and Italy. Examples of such organizations are DACOWITS (US Defense Department Advisory Committee on Women in the Services), the National Organization of Women, or the Women's Research and Education Institute at Minerva Center in the United States. The information regarding the activities of these organizations came from both primary and secondary sources. In terms of primary sources, a number of veterans, soldiers, academics and researchers working on the subject contributed to the collection of data by offering recollection of days when the legislative days took place. Many of these stories were shared with me via H-Net listserv that is administrated by Linda De Pauw Grant and MINERVA Center in Washington D.C. Dr. De Pauw asked members of the listserv on my behalf to share any information they wanted. These outstanding women and men were willing to offer their opinion regarding gender integration in the military as well as their personal stories. In Italy, I studied the radical and Marxist feminist groups Movimento Di Liberazione Di Donne and Rivolta Femminile, and the new Gruppo Donna Soldato representing both current and aspiring women soldiers, primarily through secondary sources. Although I have established contacts and spoken to a number of female soldiers, none wanted to have any of our conversations shared in public. My requests for information to Sabrina Britoni Piazza, leader of Gruppo Donna Soldato, were politely redirected to the Ministry of Defense.

Culture

Data Collection: Data on types of religion were collected from the United States Department of State Background Notes, which is available online for each state included in this study.[1] Data on all states regarding family values, morals, religiosity, and role of women in society, in politics, and in the workplace were obtained from the World Values Survey. They were based on two indicators:

    a. Whether people think that men make better executives than women.
    b. Whether people think that men make better political leaders than women.

The answers for the first indicator (a) were available for only 13 states from the study years 2005–2008, while the answers for the second indicator (b) were available from 19 states, of which 6 are from study years 1996–1999. The full survey questions and tables with answers for each state are available in Appendix C. I have simplified the analysis by combining answers 1 and 2 (Agree Strongly / Agree) and answers 3 and 4 (Disagree / Disagree Strongly) and using the mean for each as a state's score in my dataset.

    When it comes to measuring a level of religiosity, there is no scientific consensus on how exactly we are supposed to assess that. Those individuals who declare that they believe in God may not attend religious services on a regular basis, and those who do attend may be just trying to decide which religion to choose or might be under pressure to conform. Measuring it therefore requires looking into a number of dimensions that reflect people's behavior and beliefs regarding a whole set of religious, social and cultural traditions. To complete this task I have adopted the Strength of Religiosity Scale developed by Pippa Norris and Ronald Inglehart by using the data from 75 different states pooled by the World Values Survey 1981–2001.[2] This scale includes answers to the following questions:

    1. How important is God in your life? (% Very, scaled 6–10).
    2. Do you find that you get comfort and strength from religion? (% Yes).
    3. Do you believe in God? (% Yes).
    4. Independently of whether you go to church or not, would you say you are a religious person? (% Religious).

---

  1   These data are updated frequently and all of them are from years 2008 and 2009. Available at: http://www.state.gov/r/pa/ei/bgn/ (accessed January 29, 2009).

  2   Norris and Inglehart (2003b: 53–55). I have used the factor analysis of six indicators from the WVS survey that Norris and Inglehart have used to develop their 100-point Religiosity Strength Scale. Each state was assigned the score based on the answers in the survey.

5. Apart from weddings, funerals and christenings, about how often do you attend religious services these days? (% Once a week or more).
6. Do you believe in life after death? (% Yes).

The factor analysis and scores for all 24 states in this study were already available, so no additional analysis was performed.

International Security Context

Data Collection: Information regarding levels of threat for each state was analyzed by examining the Level of Conflict Intensity developed by Julian Lindley-French and Franco Algieri in the Venusberg Report. Data was adapted by using the Bertelsmann Foundation European Defense Strategy 2nd Venusberg Report published in 2004 available at: http://www.cap.uni-muenchen.de/download/2004/2004_Venusberg_Report.pdf. Each state was placed and coded on the Conflict Intensity Scale of 1 to 10, in which 1 is the lowest level of readiness and capabilities for warfare, and 10 is for the highest level of global operational readiness reserved only for the United States.

Appendix B

Questionnaire to the Committee on Women in the NATO Armed Forces Delegates

Percentage of Women in Total Active Force

1. Do women have access to military academies?

   Yes or No

2. Are there occupational restrictions?

   Many (no enlisted women) Few (submarines, special units)None

3. Are there formal rank restrictions?

   Yes or No

4. What is the total percentage of women in officers' ranks?

5. Is there training segregation?

   Total   Partial   None

6. Are there family programs (maternity, child care, paid leave)?

   None   Few   Many

Harassment Regulations (anti-discrimination regulations and monitoring within the armed services)

   None   Few   Many

Appendix C

Societal Values Variables World Values Survey Questions

D059. For each of the following statements I read out, can you tell me how much you agree with each. Do you agree strongly, agree, disagree, or disagree strongly?

On the whole, men make better political leaders than women do

Possible answers:

> 1 Agree strongly
> 2 Agree
> 3 Disagree
> 4 Strongly disagree

Belgium [1999], Bulgaria [1999], Canada [2000], Czech Republic [1999], Denmark [1999], France [1999], Germany East [1999], Germany West [1999], Great Britain [1999], Greece [1999], Hungary [1999], Italy [1999], Latvia [1999], Lithuania [1999], Luxembourg [1999], Netherlands [1999], Norway [1996], Poland [1999], Portugal [1999], Romania [1999], Slovenia [1999], Spain [1999], Spain [2000], Turkey [2001], Turkey [2001], United States [1999].

Survey years 2005–2008

Men make better business executives than women do

V63. For each of the following statements I read out, can you tell me how much you agree with each. Do you agree strongly, agree, disagree, or disagree strongly? On the whole, men make better business executives than women do.

(V63) Men make better business executives than women do

Possible answers:
1 Agree strongly
2 Agree
3 Disagree
4 Strongly disagree

Bulgaria [2006], Canada [2006], France [2006], Germany [2006], Great Britain [2006], Italy [2005], Netherlands [2006], Poland [2005], Romania [2005], Slovenia [2005], Spain [2007], Turkey [2007], United States [2006].

Appendix D

Cultural Values Variables World Values Survey Questions

A. Believe in: God

F050. Which, if any, of the following do you believe in?
God

Possible answers:

0   No
1   Yes

B. How often do you attend religious services

F028. Apart from weddings, funerals and christenings, about how often do you attend religious services these days?

Possible answers:

1. More than once a week
2. Once a week
3. Once a month
4. Only on special holy days/Christmas/Easter days
5. Other specific holy days
6. Once a year
7. Less often
8. Never practically never

Appendix E

Percentages of Women in Technical and Professional Fields

Country	Percentage of women in professional and technical fields
Belgium	49
Bulgaria	60
Canada	56
Czech Republic	52
Denmark	53
France	47
Germany	50
Greece	49
Hungary	62
Italy	46
Latvia	65
Lithuania	67
Luxembourg	N/A
Netherlands	50
Norway	50
Poland	61
Portugal	50
Romania	57
Slovakia	58
Slovenia	57
Spain	48
Turkey	32
United Kingdom	47
United States of America	56

Source: United Nations Development Program Human Development Index data.

Appendix F

Unemployment Rates in 2008

Country	Unemployment rate
Belgium	6.98
Bulgaria	5.6
Canada	6.13
Czech Republic	4.41
Denmark	3.36
France	7.8
Germany	7.31
Greece	7.65
Hungary	7.83
Italy	6.75
Latvia	7.3
Lithuania	5.7
Luxembourg	4.88
Netherlands	2.76
Norway	2.52
Poland	7.18
Portugal	7.74
Romania	4
Slovakia	9.57
Slovenia	4.5
Spain	11.38
Turkey	12
United Kingdom	5.64
United States of America	5.78

Source: UNECE Statistical Division Database, compiled from national and international (EUROSTAT, OECD, CIS) official sources.

Appendix G

Women Percentage in Lower or Single Houses after Parliamentary Renewals in 2008

Rank	Country	Lower or single House				Upper House or Senate			
		Elections	Seats*	Women	% W	Elections	Seats	Women	% W
8	Netherlands	11 2006	150	62	41.3	5 2007	75	26	34.7
9	Norway	9 2009	169	66	39.1	—	—	—	—
10	Denmark	11 2007	179	68	38.0	—	—	—	—
13	Spain	3 2008	350	127	36.3	3 2008	263	79	30.0
15	Belgium	6 2007	150	53	35.3	6 2007	71	27	38.0
18	Germany	9 2009	622	204	32.8	N.A.	69	15	21.7
33	Portugal	9 2009	230	63	27.4	—	—	—	—
47	Canada	10 2008	308	68	22.1	N.A.	93	32	34.4
51	Italy	4 2008	630	134	21.3	4 2008	322	58	18.0
54	Bulgaria	7 2009	240	50	20.8	—	—	—	—
55	Poland	10 2007	460	93	20.2	10 2007	100	8	8.0
56	Latvia	10 2006	100	20	20.0	—	—	—	—
56	Luxembourg	6 2009	60	12	20.0	—	—	—	—
58	United Kingdom	5 2005	646	126	19.5	N.A.	746	147	19.7
59	Slovakia	6 2006	150	29	19.3	—	—	—	—
64	France	6 2007	577	105	18.2	9 2008	343	75	21.9
66	Lithuania	10 2008	141	25	17.7	—	—	—	—
68	Greece	10 2009	300	52	17.3	—	—	—	—
"	United States of America	11 2008	435	73	16.8	11 2008	98	15	15.3
76	Czech Republic	6 2006	200	31	15.5	10 2008	81	14	17.3
87	Slovenia	9 2008	90	12	13.3	11 2007	40	1	2.5
95	Romania	11 2008	334	38	11.4	11 2008	137	8	5.8
97	Hungary	4 2006	386	43	11.1	—	—	—	—
105	Turkey	7 2007	549	50	9.1	—	—	—	—

Appendix H

Women in Ministerial Positions (% of positions)

HDI Rank	Country	2008
1	Norway	56
4	Canada	16
6	Netherlands	33
8	France	47
11	Luxembourg	14
13	United States	24
15	Spain	44
16	Denmark	37
17	Belgium	23
18	Italy	24
21	United Kingdom	23
22	Germany	33
25	Greece	12
29	Slovenia	18
34	Portugal	13
36	Czech Republic	13
41	Poland	26
42	Slovakia	13
43	Hungary	21
46	Lithuania	23
48	Latvia	22
61	Bulgaria	24
63	Romania	0
79	Turkey	4

Notes: Data are as of January 2008. The total includes deputy prime ministers and ministers. Prime ministers were also included when they held ministerial portfolios. Vice-presidents and heads of governmental or public agencies are not included. IPU (2009).

Appendix I

Gross National Income per Capita 2008: Purchasing Power Parity (in International Dollars)

Country	PPP GNI
Belgium	34,760
Bulgaria	11,950
Canada	36,220
Czech Republic	22,790
Denmark	37,280
France	34,400
Germany	35,940
Greece	28,470
Hungary	17,790
Italy	30,250
Latvia	16,740
Lithuania	18,210
Luxembourg	64,320
Netherlands	41,670
Norway	58,500
Poland	17,310
Portugal	22,080
Romania	13,500
Slovakia	21,300
Slovenia	26,910
Spain	31,130
Turkey	13,770
United Kingdom	36,100
United States of America	46,970

Source: World Bank 2008. Available at: http://siteresources.worldbank.org/DATASTATISTICS/Resources/GNIPC.pdf.

Bibliography

Adams-Ender, Clara, and Blair Walker. *My Rise to the Stars: How a Sharecropper's Daughter Became an Army General*. Lake Ridge, VA: Cape Associates, 2001.

Adams, Thomas. *The Army after Next: The First Postindustrial Army*. Stanford, CA: Stanford University Press, 2008.

Adler, Emanuel. "The Emergence of Cooperation: National Epistemic Communities and the International Evolution of the Idea of Nuclear Arms Control." *International Organization*. Vol. 46, no. 1 (1992): 101–145.

Ahmed, Leila. *Women and Gender in Islam: Historical Roots of a Modern Debate*. New Haven, CT: Yale University Press, 1992.

Albanese, Patritzia. *Mothers of the Nation: Women, Families and Nationalism in Twentieth Century Europe*. Toronto, CA: University of Toronto Press, 2006.

Alexandre, Laurien. "Genderizing International Studies: Revisioning Concepts and Curriculum." *International Studies Notes*. Vol. 14, no. 1 (1989): 5–8.

Alfonso, Kristal. *Femme Fatale: An Examination of the Role of Women in Combat and Future Policy Implications*. Maxwell Air Force Base, AL: Air University Press, 2009.

Alkire, Sabina "A Conceptual Framework for Human Security." Centre for Research on Inequality, Human Security, and Ethnicity (CRISE), Working Paper 2, University of Oxford, 2003.

Al-Jawaheri, Husein. *Women in Iraq: The Gender Impact of International Sanctions*. London: Zed Books, 2008.

Allen, Charles. "Lessons Not Learned: Civil-Military Disconnect in Afghanistan." *Armed Forces Journal*. Vol. 148, no. 2 (2010): 30–33.

Angrist, J. "Lifetime Earnings and the Vietnam Era Draft Lottery: Evidence from Social Security Administrative Records." *American Economic Review*. Vol. 80, no. 3 (1990): 313–336.

Antecol, Heather and Deborah Cobb Clark. "Men, Women and Sexual Harassment in the U.S. Military." *Gender Issues*. Vol. 19, no. 1 (December 2001): 3–18.

Appy, Christian G., *Working Class War: American Combat Soldiers and Vietnam*. Chapel Hill, NC: University of North Carolina Press, 1993.

Arpad, S. Susan and Sarolta Marinovich. "Why Hasn't There Been a Strong Women's Movement in Hungary?" *Journal of Popular Culture*. Vol. 29, no. 2 (1995): 77–96.

Åselius, Gunnar. "Swedish Strategic Culture after 1945." *Cooperation and Conflict*. Vol. 40, no. 1 (2005): 25–44.

Ash, J. Colin K., Bernard Udis, and Robert F. McNown, "Enlistment in the All-Volunteer Force: A Military Personnel Supply Model and Its Forecasts." *The American Economic Review*. Vol. 73, no. 1 (1983): 145–155.

Ashworth, Lucian, and Larry Swatuck. "Masculinity and the Fear of Emasculation in International Relations Theory." In Marysia Zalewski and Jane Parpart (eds.), *The "Man" Question in International Relations*. Boulder, CO: Westview Press, 1998.

Avant, Deborah. *The Market for Force: The Consequences of Privatizing Security*. Cambridge: Cambridge University Press, 2005.

Baldor, Lolita C. "Army More Selective on Recruits, Re-Enlistments." May 22, 2012. Available at: http://www.armytimes.com/news/2012/05/ap-army-more-selective-recruits-reenlistments-052212/ (accessed November 5, 2012).

Balzacq, Thierry. *Securitization Theory: How Security Problems Emerge and Dissolve*. Milton Park, Abingdon: Routledge, ٢٠١١.

Barak-Erez, Daphne. "The Feminist Battle for Citizenship: Between Combat Duties and Conscientious Objection." *Cardozo Journal of Law & Gender*. Vol. 13 (2007): 531–560.

Barrett, F.J. "The Organisational Construction of Hegemonic Masculinity: The Case of the US Navy." *Gender, Work and Organization*. Vol. 3, no. 3 (1996): 129–142.

Baucom, Donald R. "The Professional Soldier and the Warrior Spirit." *Strategic Review*. Vol. 13 (1985): 57–66.

Beall, J. "Trickle Down or Rising Tide? Lessons on Mainstreaming Gender Policy from Columbia and South Africa." *Social Policy and Administration*. Vol. 32, no. 5 (1998): 513–534.

Behnke, A. "Presence and Creation: A Few (meta-)Critical Comments on the C.A.S.E. Manifesto." *Security Dialogue*. Vol. 38, no. 1 (2007): 105–111.

Békés, Csaba, Malcolm Byme, and János Rainer. *The 1956 Hungarian Revolution: A History in Documents*. National Security Archive Cold War Readers. New York: Central European University Press, 2002.

Belkin, Aaron. *Bring Me Men: Military Masculinity and the Benign Facade of American Empire, 1899-2012*. New York: Columbia University Press, 2012.

Benedict, Helen. "For Women Warriors, Deep Wounds, Little Care." *New York Times*, May 26, 2008, Op-ed.

Benedict, Helen. *Lonely Soldier: The Private War of Women Serving in Iraq*. Boston, MA: Beacon Press, 2010.

Berkovitch, Nitza. *From Motherhood to Citizenship: Women's Rights and International Organizations*. Baltimore, MD: Johns Hopkins University Press, 1999.

Berryman, Sue E. *Who Serves? The Persistent Myth of the Underclass Army*. Boulder, CO: Westview, 1988.

Binkin, Martin. *Who Will Fight the Next War? The Changing Face of the American Military*. Washington, DC: Brookings Institution, 1993.

Binkin, Martin, and Shirley Bach. *Women and the Military*. Washington, DC: Brooking Institute, 1977.

Birke, Lynda. *Women, Feminism and Biology: The Feminist Challenge*. London: Wheatsheaf Books, 1986.

Blair Commission. *Congressional Commission on Military Training and Gender-Related Issues: Final Report*. 4 vols. Washington, DC: U.S. Department of Defense, 1999.

Blanchard, Eric. "Gender, International Relations, and the Development of Feminist Security Theory." *Journal of Women in Culture and Society*. Vol. 28, no. 4 (2003): 1289–1313.

Bland, Douglas. "Patterns in Liberal Democratic Civil-Military Relations." *Armed Forces and Society*. Vol. 27, no. 4 (2001): 525–540.

Blanton, DeAnne, and Lauren Cook Wike. *They Fought Like Demons: Women Soldiers in the Civil War*. Baton Rouge, LA: LSU Press, 2002.

Boldizsár, Iván (ed.). *Hungary*. Budapest: Corvina Press, 1965.

Bolgar, Judit. "Women in the Hungarian Armed Forces." *Minerva: Quarterly Report on Women and the Military*. Vol. 17 (1999): 92–99.

Book, Elizabeth. "Military Women 200,000 and Counting," *National Defense Magazine*. October 2001. Available at: http://www.nationaldefensemagazine. org/ARCHIVE/2001/OCTOBER/Pages/Military_Women4193.aspx (accessed June 25, 2009).

Booth, Ken Russell, and B. Trood. *Strategic Cultures in the Asia-Pacific Region*. New York: St. Martin's Press, 1999.

Bragg, Janet. *Soaring above Setbacks: The Autobiography of Janet Harmon Bragg, African American Aviator*. Washington, DC: Smithsonian Institution Press, 1996.

Bragg, Rick. *Jessica Lynch Story: I am a Soldier Too*. New York: Knop Publishing Group, 2003.

Breuer, William B. *War and American Women: Heroism, Deeds, and Controversy*. Westport, CT: Praeger, 1997.

Brown, Charles. "Military Enlistments: What Can We Learn From Geographic Variation?" *American Economic Review*. Vol. 75, No. 1 (1985): 228–234.

Burant, Stephen R. *Hungary: A Country Study*. Washington, DC: Federal Research Division, Library of Congress Research September, 1989.

Burelli, David. "Women in Combat: Issues for Congress." Washington, DC: Congressional Research Service, 2013. Available at: http://www.fas.org/ sgp/crs/natsec/R42075.pdf. (accessed May 17, 2013).

Buzan, Barry. *People, States and Fear: The National Security Problem in International Relations*. Hemel Hempstead: Harvester Wheatsheaf, 1983.

Buzan, Barry, Ole Wæver, and Jaap de Wilde. *Security: A New Framework for Analysis*. Boulder, CO: Lynne Rienner, 1998.

Caforio, Giuseppe. *Handbook of the Sociology of the Military*. New York: Kluwer Academic/Plenum, 2003.

Calahan, Philip D. *The Code of the Warrior: The Kinder, Gentler Military and Marksmanship: Changing a Culture.* Strategy Research Project. Carlisle Barracks, PA: U.S. Army War College, 2002. Available at: http://handle.dtic. mil/100.2/ADA401746 (accessed May 2010).

Campbell, D'Ann. "Women in Combat: The World War II Experience in the United States, Great Britain, Germany, and the Soviet Union." *Journal of Military History.* Vol. 57 (1993): 301–323.

Canosa, Romano, *Il Giudice e la donna. Cento anni di sentenze sulla condizione femminile in Italia.* Milano: Mazzotta, 1978.

Caprioli, Mary. "Feminist IR Theory and Quantitative Analysis: A Critical Analysis." *International Studies Review.* Vol. 6, no. 2 (2004): 253–269.

Career Progression of Active Duty Career Women Army Officers. Carlisle Barracks, PA: Army War College, 1996.

Carpenter, Charli. "Gender Theory in World Politics: Contributions of a Non-Feminist Standpoint." *International Studies Review.* Vol. 5, no. 3 (2002): 153–165.

Carpenter, Charli. "'Women and Children First': Gender, Norms, and Humanitarian Evacuation in the Balkans." *International Organization.* Vol. 57, no. 4 (2003): 428–477.

Carpenter, Charli. "Women, Children, and Other Vulnerable Groups: Gender, Strategic Frames, and the Protection of Civilians as a Transnational Issue." *International Studies Quarterly.* Vol. 49, no. 2 (2005): 295–335.

Carreiras, Helena. "The Role of Women in the Armed Forces of NATO Countries: Military Constraints and Professional Identities." *Minerva: Quarterly Report on Women and the Military.* Vol. 17, no. 3–4 (1999): 46–57. Available at: http:// findarticles.com/p/articles/mi_m0EXI/is_1999_Fall-Winter/ai_66239866/ (accessed July 13, 2009).

Carreiras, Helena. *Gender and the Military: Women in the Armed Forces of Western Democracies.* London: Routledge, 2006.

Carroll, Berenice and Barbara W. Hall. "Feminist Perspectives Women and the Use of Force." In Ruth H. Howes and Michael R. Stevenson (eds.), *Women and the Use of Military Force,* 11–22. Boulder, CO: Lynne Rienner, 1993.

Carroll, Susan. *The Impact of Women in Public Office.* Bloomington, IN: University of Indiana Press, 2001.

Carter, Phillip. "War Dames: American Female Soldiers to Fight in Iraq." *The Washington Monthly.* Vol. 34 (2002): 32–37.

Caul, Miki. "Women's Representation in Parliament: The Role of Political Parties." *Party Politics.* Vol. 5, no. 1 (1999): 79–98.

Cavallaro, Gina. "Assault Prevention Officer Faces House Panel." *Army Times,* September 12, 2008. Available at: http://www.armytimes.com/news/2008/09/ army_whitleyhearing_091108w/ (accessed July 20, 2009).

Cavarero, Adriana. "Per una teoria della differenza sessuale." In various authors, *Diotima. Il pensiero della differenza sessuale,* 43–79. Milan: La Tartaruga, 1987.

Cavarero, Adriana. "Il pensiero femministe. Un approccio teoretico." In Adriana Cavarero and Franco Restaino (eds.), *Le filosofie femministe*, 111–164.Turin: Paravia, 1999.

Cave, Damien. "A Combat Role, and Anguish, Too, Women at Arms." *New York Times*, October 31, 2009. Available at: http://www.nytimes.com/2009/11/01/us/01trauma.htmL (accessed May 25, 2012).

Cerny, Phil. "'Iron Triangle' to 'Golden Pentangles'?: Globalizing the Policy Process." *Global Governance*. Vol. 7 (2001): 397–410.

Center for Military Readiness. *Issues: Recruiting/The Draft: Court Dismisses Lawsuit to Include Women in Draft Registration*. September 4, 2003. Available at: http://cmrlink.org/recruit.asp?DocID=212. (accessed January 23, 2009).

Center for Military Readiness. *Issues: Women in Combat: Bush Administration Upholds Law and Regulations Exempting Women from Land Combat.* May 31, 2002. Available at: http://cmrlink.org/WomenInCombat.asp?docID=154. (accessed January 23, 2009).

Chapkis, Wendy. *Loaded Questions: Women in Militaries*. Washington, DC: Institute for Policy Studies, 1982.

Chema, J.R. "Arresting 'Tailhook': The Prosecution of Sexual Harassment in the Military." *Military Review*. Vol. 140 (Spring 1993): 1–64.

Chen, L., J. Leaning, and V. Narasimhan. *Global Health Challenges for Human Security*. Cambridge, MA: Harvard University Press, 2004.

Cheng, C. "Marginalised Masculinities and Hegemonic Masculinity: An Introduction." *The Journal of Men's Studies*. Vol. 7, no. 13 (1999): 295–314.

Chiavoloa Birnbaum, Lucia. *Liberazione della donna: Feminism in Italy.* Middletown, CT: Wesleyan University Press, 1986.

Churchill, Janet. *On Wings to War: Teresa James, Aviator*. Manhattan, KS: Sunflower University Press, 1992.

Cillers, Jakkie, and Lindy Heinecken. *South Africa: Emerging from a Time Warp: The Role of Women*. New York: Oxford University Press, 2000.

Clark Sander, Georgia. *Women in Combat: The US Military and the Impact of the Persian Gulf.* Westport, CT: Bergin and Garvey, 1997.

Claude, Inis. *Swords into Plowshares: The Problems and Progress of International Organization*. New York: Random House, 1971.

Clausewitz, Carl. *On War*. London: N. Trübner, 1873.

Clayton, Ellen. "Female Warriors: Memorials of Female Valour and Heroism." In E. Clayton, *Mythological Ages to the Present Era*. 2 vols. London: Tinsley Brothers, 1879.

Clayton, Roberts. *The Logic of Historical Explanation*. University Park, PA: Pennsylvania State University Press, 1996.

Cockburn, Cynthia, and Zarkov Dubravaka. *The Post-War Moment: Militaries, Masculinities, and International Peacekeeping*. London: Zed Books, 2002.

Cohen, E.A. *Citizens and Soldiers: The Dilemmas of Military Service*. Ithaca, NY: Cornell University Press, 1985.

Cohen, Miriam. *Workshop to Office: Two Generations of Italian Women in New York City, 1900–1950*. Cornell, NY: Cornell University Press, 1993.

Cohen, Stuart. *Democratic Societies and Their Armed Forces: Israel in Comparative Context*. Portland, OR: Frank Cass, 2000.

Cohn, Carol. "Slick 'ems, Glick 'ems, Christmas Trees, and Cookie Cutters: Nuclear Language and How We Learned to Pat the Bomb." *Bulletin of the Atomic Scientists*, Vol. 43 (1987): 17–24.

Cohn, Carol. "Wars, Wimps, and Women: Talking Gender and Thinking War." In Miriam Cooke and Anjela Woollacott (eds.), *Gendering War Talk*, 227–246. Princeton, NJ: Princeton University Press, 1993.

Cohn, Carol. "How Can She Claim Equal Rights When She Doesn't Have to Do as Many Push Ups as I Do? The Framing of Men's Opposition to Women's Equality in the Military." *Men and Masculinities.* Vol. 3, no. 2 (2000): 131–151.

Cohn, Carol, and Sara Ruddick. *A Feminist Ethical Perspective on Weapons of Mass Destruction*. Princeton, NJ: Princeton University Press, 2002.

Collier, Ellen. *Congressional Research Service Brief: Women in the Military*. Washington, DC: Library of Congress, 1991.

Commission on Human Security. "Feminist International Relations: Old Debates and New Direction." Last modified 2004. Available at: http://www. watsoninstitute.org/bjwa/archive/10.2/Feminist Theory/Wibben.pdf (accessed February 17, 2011).

Commission on Human Security. "Human Security Now: Report of the Commission on Human Security." Last modified 2007. Available at: http://www.nato.int/ issues/women_nato/cwinf_guidance.pdf (accessed September 2012).

Congress, House, House Armed Service Committee, Subcommittee on Personnel and Readiness, and the Defense Policy Panel. Hearings into Gender Discrimination in the Military. 102nd cong. 2nd.Assess. July 29–30, 1992.

Congress, House, House Armed Services Committee. "Women in the Military: The Tailhook Affair and the Problem of Sexual Harassment." Draft report prepared by Les Aspin and Beverly B. Byron, September 14, 1992.

Connell, Raewyn and James W. Messerschmidt. 2005. "Hegemonic Masculinity: Rethinking the Concept." *Gender and Society*, Vol. 19, no. 6 (2005): 829–859.

Conover, Pamela, and Virginia Sapiro. "Gender, Feminist Consciousness, and War." *American Journal of Political Science.* Vol. 37, no. 4 (1993): 1079–1099.

Conze, Susanne, and Fieseler Beate. *Soviet Women as Comrades-in-Army: A Blind Spot in the History of War*. Urbana, IL: University of Illinois Press, 2000.

Cooper, Richard V.L. *Military Manpower and the All-Volunteer Force*. Santa Monica, CA: Rand, 1977.

Coppola, Nicholas, and Kevin LaFrance. "The Female Infantryman: A Possibility." *Military Review*. Vol. 82 (2002): 61–66.

Corbett, A.J. *Women in Combat: The Case for Combat Exclusion (Final Report)*. Newport, RI: Naval War College, 1993.

Cornum, Rhonda, and Peter Copeland. *She Went to War: The Rhonda Cornum Story*. Novato, CA: Presidio, 1992.

Cottam, K. Jean. "Soviet Women in Combat in World War II: The Ground Forces and the Navy." *International Journal of Women's Studies.* Vol. 3, no. 4 (1980): 345–57.

Cottam, K. Jean. *Soviet Airwomen in Combat in World War II.* Manhattan, KS: Military Affairs/Aerospace Historian Publishing, 1983.

Cresswell, J.W. *Research Design: Qualitative, Quantitative and Mixed Approaches.* Thousand Oaks, CA: Sage Publications, 2003.

Cresswell, J.W, L. Vicki, and Clark Plano. *Designing and Conducting Mixed Methods Research.* Thousand Oaks, CA: Sage Publications, 2006.Culberston, Amy, Paul Rosenfeld, and Carol Newell. *Sexual Harassment in the Active Duty Navy: Findings.* Navy-Wide Survey. Washington, DC: Naval Personnel Research & Development Center, 1993.

Dahlerup, Drude. "From a Small to a Large Minority: Women in Scandinavian Politics." *Scandinavian Political Studies.* Vol. 11 no. 4 (1988): 275–97.

Daks, Brian. "Women In Combat Debate Rekindled: Killing, Wounding of 14 Female Troops Last Week Spotlights Danger." CBS News, June 30, 2005. Available at: http://www.cbsnews.com/stories/2005/06/30/earlyshow/main705317_page2.shtml?tag=contentMain;contentBody (accessed May 30, 2009).

Dale, Charles and Gilroy, Curtis. "Enlistments in the All-Volunteer Force: Note." *American Economic Review.* Vol. 75, no. 3 (1985): 547–551.

D'Amico, Francine. *Critical Feminism: Deconstructing Gender, Nationalism, & War.* Boulder, CO: Lynne Rienner, 2006.

D'Amico, Francine, and Peter Beckman. *Women in World Politics: An Introduction.* Westport, CT: Bergin and Garvey, 1995.

Dandecker, Christopher, and Mady Wechsller Segal. "Gender Integration in Armed Forces: Recent Policy Developments in the United Kingdom." *Armed Forces & Society.* Vol. 23, no. 1 (1996): 29–47.

Deck-Partyka, Alicja. *Poland: A Unique Country and Its People.* Bloomington, IN: AuthorHouse, 2006.

De Clementi, A. "The Feminist Movement in Italy." In R. Braidotti and G. Griffin (eds.), *Thinking Differently: A Reader in European Women's Studies,* 332–340. New York: Zed Books, 2002.

De Grazia, Victoria. *How Fascism Ruled Women (1922–1945).* Berkeley, CA: University of California Press, 1993.

De Grand, Alexander. "Women under Italian Fascism." *The Historical Journal.* Vol. 19, no. 4 (1976): 947–968.

DeGroot, Gerard. *Gender Stereotypes, the Military, and Peacekeeping.* London: Frank Cass, 2001.

DeGroot, Gerard, and Corinna Peniston-Bird. *A Soldier and a Woman: Sexual Integration in the Military.* New York: Longman, 2000.

Department of the Army. *Secretary of the Army's Senior Review Panel Report on Sexual Harassment.* Vol. 2: *Data Report.* Washington, DC: Department of the Army, 1997.

Department of the Army. Army Regulation 600–613, Personnel—General Army Policy for the Assignment of Female Soldiers. Washington, DC, March 27, 1992. Available at: http://www.apd.army.mil/pdffiles/r600_13.pdf (accessed April 17, 2012).

Department of the Army Inspector General. *Special Investigation of Initial Entry Training, Equal Opportunity and Sexual Harassment Policies and Procedures December 1996–April 1997. July 22, 1997. Human Relations Action Plan: The Human Dimensions of Combat Readiness*. Department of the Army, July 1997.

De Pauw, Linda. "Women in Combat: The Revolutionary War Experience." *Armed Forces and Society*. Vol. 7, no. 2 (1981): 209–226.

De Pauw, Linda Grant. *Battle Cries and Lullabies: Women in War from Prehistory to the Present*. Norman, OK: University of Oklahoma Press, 1998.

Devilbiss, Margaret Conrad. "Gender Integration and Unit Deployment: A Study of GI Jo." *Armed Forces and Society*. Vol. 11, no. 4 (1985): 523–552.

Devilbiss, Margaret Conrad. *Women and Military Service*. Maxwell Air Force Base, AL: Air University Press, 1990.

Di Leonardo, Micaela. "Morals, Mothers, and Militarism: Antimilitarism and Feminist Theory." *Feminist Studies*. Vol. 11, no. 3 (1985): 599–617.

Disler, Edith. "The Feminine as the Force Multiplier." In James E. Parco and David A. Levy, (eds.), *Attitudes Aren't Free: Thinking Deeply About Diversity in the U.S. Armed Forces,* 363–379. Maxwell Air Force Base, AL: Air University Press, 2010.

Dodds, F., and T. Pippard. *Human and Environmental Security: An Agenda for Change*. London: Earthscan, 2005.

Dolan, Kathleen, and Lynne Ford. *Are All Women Legislators Alike?* In Sue Thomas and Clyde Wilcox, (eds.), *Women and Elective Office: Past, Present and Future*, 73–86. New York: Oxford University Press, 1998.

Dollar, David, and Roberta Gatti. "Gender Inequality, Income and Growth: Are Good Times Good for Women?" The World Bank, Policy Research Report on Gender and Development, Working Paper Series, no. 1 (1999). Available at: http://siteresources.worldbank.org/INTGENDER/Resources/wp1.pdf (accessed September 5, 2011).

Donnelly, Elaine. "Social Experimentation in the Military." Speech given at the Heritage Foundation on March 3, 1995. Transcripts available at: http://www.heritage.org/research/nationalsecurity/upload/92136_1.pdf (accessed May 25, 2009).

Donnelly, Elaine. "Defending the Culture of the Military." In E. Donnelly, *Attitudes Aren't Free: Thinking Deeply about Diversity in the U.S. Armed Forces*, 249–292. Maxwell Air Force Base, AL: Air University Press, 2010.

Dowler, Lorraine. "Women on the Frontlines: Rethinking war Narratives post 9/11." *GeoJournal*. 58 (2002): 159–165.

Dudink, Stefan, Karen Hagemann, and John Tosh. *Masculinities in Politics and War Gendering Modern History*. Manchester: Manchester University Press, 2004.

Dunay, Pal. "The Half-Hearted Transformation of Hungarian Military." *European Security*. Vol. 14, no. 1 (2005): 17–32.

Dunne, Tim, Milja Kurki, and Steve Smith (eds). *International Relations Theories: Discipline and Diversity*. Oxford: Oxford University Press, 2013.

Eberhardt, Eva. "Situation of Women in Hungary." *Commission of the European Communities*. Vol. 32 (1991): 153–163.

Eckstein, Harry A. "Culturalist Theory of Political Change." *The American Political Science Review*. Vol. 82, no. 3 (1988): 789–804.

Edmonds, Sarah Emma. *Memoirs of a Soldier, Nurse, and Spy: A Woman's Adventures in the Union Army*. DeKalb, IL: Northern Illinois University Press, 1999.

Edmunds, Timothy, and Malesic Markan. *Defense Transformation in Europe: Evolving Military Roles*. Fairfax, VA: IOA Press, 2005.

Einhorn, Barbara. *Cinderella Goes to Market: Citizenship, Gender and Women's Movements in East Central Europe*. London: Verso, 1993.

El-Bushra, Judy. *Transforming Conflict: Some Thoughts on a Gendered Understanding of Conflict Processes*. London: Zed Books, 2000.

Elder, Glen, H. Lin Wang, Naomi J. Spence, Daniel E. Adkins, and Tyson H. Brown. "Pathways to the All-Volunteer Military." *Social Science Quarterly*. Vol. 91, no. 2 (2010): 455–475.

Ellefson, K.G. *Advancing Army Women as Senior Leaders: Understanding the Obstacles*. Carlisle Barracks, PA: Army War College, 1998.

Elshtain, Jean Bethke. *Public Man, Private Woman: Women in Social and Political Thought*. Oxford, OH: Martin Robertson, 1981.

Elshtain, Jean Bethke. "Reflections on War and Political Discourse: Realism, Just War, and Feminism in the Nuclear Age." *Political Theory*. Vol. 13, no. 1 (1985): 39–57.

Elshtain, Jean Bethke. *Women in War*. New York: Basic Books, 1987.

Elshtain, Jean Bethke, and Shelia Tobias. *Women, Militarism, and War*. Savage, MD: Rowman and Littlefield, 1990.Enloe, Cynthia. *Does Khaki Become You?* Boston, MA: South End Press, 1983.

Enloe, Cynthia. "Feminists Thinking about War, Militarism, and Peace." In Beth B. Hess and Myra Marx Ferree (eds.), *Analyzing Gender: A Handbook* of *Social Science Research*, 526–47. Newbury Park, CA: Sage, 1987.

Enloe, Cynthia. "Women and Children: Making Feminist Sense of the Persian Gulf Crisis." *Village Voice*, September 25, 1990a.

Enloe, Cynthia. *Bananas, Beaches, and Bases: Making Feminist Sense of International Politics*. Berkeley, CA: University of California Press, 1990b.

Enloe, Cynthia. *Maneuvers: The International Politics of Militarizing Women's Lives*. Berkeley, CA: University of California Press, 2000.

Ergas, Yasmine. "Feminism in the Italian Party System: Women's Politics in a Decade of Turmoil." *Comparative Politics*. Vol. 14, no. 3 (1982): 253–279.

Eulriet, Irene. *Women and the Military in Europe: Comparing Public Cultures*. Basingstoke: Palgrave Macmillan, 2012.

Fabian, Katalin. *Contemporary Women's Movements in Hungary: Globalization, Democracy, and Gender Equality*. Baltimore, MD: Johns Hopkins University Press, 2009.

Faludi, Susan. *The Terror Dream: Fear and Fantasy in Post-9/11 America.* New York: Meteropolitan Books, 2007.

Fareed, Zakaria, "Realism and Domestic Politics: A Review Essay." *International Security*. Vol. 17, no. 1 (1992): 190–196.

Fátima, Mernissi. *The Veil and the Male Elite: A Feminist Interpretation of Women's Rights in Islam*. Bloomington, IN: University of Indiana Press, 1991.

Feinman, Ilene Rose. *Citizenship Rites: Feminist Soldiers and Feminist Anti-Militarists*. New York: New York University Press, 2000.

Fenner, Lorry M. and Marie E. de Young. *Women in Combat: Civic Duty or Military Liability?* Washington, DC: Georgetown University Press, 2001.

Finch, M. "Women in Combat: One Commissioner Reports." *Minerva: Quarterly Report on Women and the Military*. Vol. 12, no. 1 (1994): 1–12.

Finnemore, Martha. "International Organizations as Teachers of Norms: The United Nations' Educational, Scientific, and Cultural Organization and Science Policy." *International Organization*. Vol. 47, no. 4 (1993): 565–597.

Finnemore Martha and Kathryn Sikkink. "International Norms and Political Change." *International Organization*. Vol. 52, no. 4 (1998): 887–917.

Fischer, Hannah. *American War and Military Operations Casualties: Lists and Statistics*. Washington, DC: Department of the Navy, Navy Historical Center, 2005. Available at: http://www.history.navy.mil/library/online/american%20war%20casualty.htm (accessed November 2012).

Floyd, Rita. "Human Security and the Copenhagen School's Securitization Approach: Conceptualizing Human Security as a Securitizing Move." *Human Security Journal*. Vol. 5 (2007): 38–49.

Flynn, George Q. *The Draft, 1940–1973*. Lawrence, KS: University Press of Kansas, 1993.

Flynn, George Q. *Conscription and Democracy: The Draft in France, Great Britain, and the United States*. Westport, CT: Greenwood Press, 2002.Fortin, Noonie. *Potpourri of War*. San Antonio, TX: LangMark Publishing, 1998.

Fowler, Tillie K. "Report of the Panel to Review Sexual Misconduct Allegations at the U.S. Air Force Academy." Available at: http://www.defense.gov/news/sep2003/d20030922usafareport.pdf (accessed March 20, 2013).

Fraser-Andrews, L.J. *Women in Combat: The Operational Impact of Meeting a National Security Necessity (Final Report)*. Newport, RI: Naval War College, 1991.

Freeman, Carla. "Is Local:Global as Feminine:Masculine? Rethinking the Gender of Globalization." *Signs: Journal of Women in Culture and Society*. Vol. 26, no. 4 (2001): 1007–1037.

Fuszara, Marlgorata "Between Feminism and the Catholic Church." *Czech Sociological Review*. Vol. 41, no. 6 (2005): 1057–1075.

Gailey, Christine Ward. "Women and Warfare: Shifting Status in Precapitalist State Formation." *Culture*. Vol. 4, (1984): 61–70.

Galagher, Nancy. *Gender Gap in Popular Attitudes toward the Use of Force*. Boulder, CO: Lynne Rienner, 1993.

Gandy, Kim. "Below the Belt: Stop Rape and Assault: And That's an Order." 6 April, 2009. Available at: http://www.now.org/news/note/040609. html (accessed June 4, 2009).

Garcia, Federico, Kletus S. Lawler and David L. Reese. "Women at Sea: Unplanned Losses & Accession Planning." Alexandria, VA: Center for Naval Analysis, CRM-98, March 1999.

Gardam, Judith. "Gender and Non-Combatant Immunity." *Transnational Law and Contemporary Problems*. Vol. 3 (1993): 345–370.

Gat, Azar. "Female Participation in War: Bio-Cultural Interactions." *Journal of Strategic Studies*. Vol. 23 (2000): 21–31.

Gavin, Lettie. *American Women in World War I: They Also Served*. Niwot, CO: University Press of Colorado, 1997.

Gelb, Joyce. *Gender Policies in Japan and the United States: Comparing Women's Movements, Rights and Policies*. New York: Palgrave Macmillan, 2003.

Gibson, Ian. "Human Security: A Framework for Peace Constructs, Gendered Perspectives and Cosmopolitan Security." *Journal of Peace, Conflict and Development*. Vol. 17 (2011): 8–101.

Gilroy, Curtis and Cindy Williams (eds.). *Service to Country: Personnel Policy and the Transformation of Western Militaries*. Cambridge, MA: JFK School of Government, 2006.

Glenn, John, Darryl Howlett, and Stuart Poore. *Neorealism versus Strategic Culture*. Burlington, VT: Ashgate Publishing Company, 2004.

Godson, Susan H. *Serving Proudly: A History of Women in the U.S. Navy*. Annapolis, MD: Naval Institute Press, 2001.

Goetz, Anne. "Mainstreaming Gender Equality to National Development Planning." In Carol Miller and Shahra Razavi (eds.), *Missionaries and Mandarins: Feminist Engagement with Development Institutions*, 42–86. London: Intermediate Tech Pubs with UNRISD, 1998.

Golding, Susan. "Women: Ready for the Challenges of the Future U.S. Armed Forces." Carlisle Barracks, PA: U.S. Army War College, 2002.

Goldstein, Joshua. *War and Gender: How Gender Shapes the War System and Vice Versa*. Cambridge: Cambridge University Press, 2001.

Goldstein, Joshua. "John Wayne and GI Jane." *Christian Science Monitor*, January 10, 2002, 11.

Goldstein, Joshua. Personal communication to author, July 27, 2009.

Goodhard, Arthur. *Poland and the Minority Races*. New York: Bretano's, 1920.

Gori, Gigliola. *Italian Fascism and the Female Body: Submissive Women and Strong Mothers*. New York: Routledge, 2004.

Gornick, Janet. C., Marcia. K. Meyers, and Katherin E. Ross. "Public Policies and the Employment of Mothers: A Cross-National Study." *Social Science Quarterly.* Vol. 79, no. 1 (1998): 35–54.

Graf, Mercedes. "Women Nurses in Spanish-American War." *Minerva: Quarterly Report on Women and the Military.* Vol. 19, no. 1 (2001): 1–24.

Grager, Nina, and Leira Halvard. "Norwegian Strategic Culture after World War II from a Local to a Global Perspective." *Cooperation and Conflict.* Vol. 40, no. 1 (2005): 45–66.

Grant, Rebecca, and Kathleen Newland. *Gender and International Relations.* Bloomington, IN: University of Indiana Press, 1991.

Gray, Colin. S. *The Geopolitics of Superpower.* Lexington, KY: University Press of Kentucky, 1988.

Gray, Colin. "Strategic Culture as Context: The First Generation of Theory Strikes Back." *Review of International Studies.* Vol. 25, no. 1 (1999): 49–69.

Grey, Sandra. "Does Size Matter? Critical Mass and Women MPs in the New Zealand House of Representatives." Paper prepared for the 51st Political Studies Association Conference, Manchester, United Kingdom, April 10–12, 2001. Available at: http://www.capwip.org/readingroom/nz_wip.pdf (accessed October 25, 2010).

Grieco, Joseph. "Anarchy and the Limits of Cooperation: A Realist Critique of the Newest Liberal Institutionalism." *International Organization.* Vol. 42, no. 3 (1988): 485–508.

Gruhzit-Hoyt, Olga. *They Also Served: American Women in World War II.* Secaucus, NJ: Carol Publishing Group, 1995.

Guadagnini, Marila. "A Partitocrazia without Women: The Case of the Italian Party System." In Joni Lovenduski and Pippa Norris (eds.), *Gender and Party Politics,* 168–204. London: Sage Publications, 1994.

Guadagnini, Marila. "The Latecomers: Italy's Equal Status and Equal Opportunity Agencies." In Dorothy M. Stetson and Amy Mazur (eds.), *Comparative State Feminism,* 150–167. London: Sage Publications, 1995.

Gudorf, Christine E. "Women and Catholic Church Politics in Eastern Europe." *Journal of Feminist Studies in Religion.* Vol. 11, no. 2 (1995): 101–116.

Gutmann, Stephanie. *The Kinder, Gentler Military: Can America's Gender-Neutral Fighting Force Still Win Wars?* New York: Scribner, 2000.

Haas, Peter. "Introduction: Epistemic Communities and International Policy Coordination." *International Organization.* Vol. 46, no. 1 (1992): 1–35.

Halliday, Fred. *Hidden from International Relations: Women and the International Arena.* Bloomington, IN: University of Indiana Press, 1991.

Halsaa, Beatrice. "A Strategic Partnership for Women's Policies in Norway." In Geertje Lycklama à Nijeholt, Virginia Vargas, and Saskia Wieringa (eds.), *Women's Movements and Public Policy in Europe, Latin America, and the Caribbean,* 167–187. New York: Garland, 1998.

Haltiner, Karl. "Decline of European Mass Armies." In Giuseppe Caforio (ed.), *Handbook of the Sociology of the Military,* 361–384. New York: Springer, 2006.

Haltiner, K, and P. Klein. "The European Post-Cold War Military Reforms and Their Impact on Civil-Military Relations." In F. Kernic, *P. Klein* and *K. Haltiner* (eds.), *The European Armed Forces in Transition: A Comparative Analysis*, 9–30. Frankfurt am Main: Peter Lang, 2005.

Hansen, Lena. "The Little Mermaid's Silent Security Dilemma and the Absence of Gender in the Copenhagen School." *Millennium*. Vol. 29, no. 2 (2000): 289–306.

Hansen, Lena. "Bosnia and the Construction of Security." *International Feminist Journal of Politics*. Vol. 3, no. 1 (2001): 55–75.

Hansen, Amanda F. and Mady W. Segal. "Value Rationales in Policy Debates on Women in the Military: A Content Analysis of Congressional Testimony, 1941–1985." *Social Science Quarterly,* Vol. 73, no. 2 (1992): 296–309.

Harding, Sandra. *The Science Question in Feminism*. Ithaca, NY: Cornell University Press, 1986.

Harding, Sandra. *Feminism and Methodology*. Bloomington, IN: University of Indiana Press, 1987.Harding, Sandra. *Whose Science: Whose Knowledge?* Ithaca, NY: Cornell University Press, 1991.

Harley, Keith. "The British Experience with an All-Volunteer Force." In Curtis Gilroy and Cindy Williams (eds.), *Service to Country: Personnel Policy and the Transformation of Western Militaries*, 287–312. Cambridge, MA: JFK School of Government. 2006.

Harper, Judith. *Women during the Civil War: An Encyclopedia*. New York: Routledge, 2004.

Harrell, Margaret. *Assessing the Assignment Policy for Army Women, RAND Corporation*. Santa Monica, CA: National Defense Research Institute, 2007.

Harrell, M.C. and Miller, L.L. *New Opportunities for Military Women: Effects upon Readiness, Cohesion, and Morale*. Santa Monica, CA: The RAND Corporation, 1997.Harries-Jenkins, Gwyn. "Women in Extended Roles in the Military: Legal Issues." *Current Sociology*. Vol. 50, no. 5 (2002): 745–769.

Hastings, David A. "From Human Development to Human Security: A Prototype Human Security Index." United Nations Economic and Social Commission for Asia and the Pacific, Working Paper WP/09/03, 2009. Available at: http://www.unescap.org/publications/detail.asp?id=1345 (accessed April 25, 2013).

Hauser, William. *America's Army in Crisis: A Study in Civil-Military Relations*. Baltimore, MD: Johns Hopkins University Press, 1973.

Hay, M.S. and C.G. Middlestead. *Women in Combat: An Overview of the Implications for Recruiting* (ARI Research Report 1568). Alexandria, VA: U.S. Army Research Institute for the Behavioral and Social Sciences, 1990.

Hays, Peter, Brenda Vallance, and Alan Van Tassel. *American Defense Policy*. Baltimore, MD: Johns Hopkins University Press, 1997.

Heidenheimer, Arnold J., Carolyn Teich Adams, and Hugh Heclo. *Comparative Public Policy: The Politics of Social Choice in Europe and America*. New York: St. Martin's Press, 1983.

Henkel, Louis O. *The Impact of Army Transformation on the Integration of Enlisted Women.* Strategy Research Project. Carlisle Barracks, PA: U.S. Army War College, 2003. (AD-A415–334) Available at: http://handle.dtic.mil/100.2/ ADA415334 (accessed June 2009).

Henry, Ryan. *Transforming the U.S. Global Defense Posture.* Newport, RI: Naval War College, 2006.

Herbert, Bob. "The Great Shame." *New York Times* online, March 20, 2009. Available at: http://www.nytimes.com/2009/03/21/opinion/21herbert.html?_ r=1 (accessed March 21, 2009).

Herbert, M.S. "Crinoline to Camouflage: Initial Entry Training and the Marginalization of Women in the Military." *Minerva: Quarterly Report on Women and the Military.* Vol. 11 (Spring 1993): 41–57.

Herbert, Melissa. *Camouflage Isn't Only for Combat: Gender, Sexuality and Women in the Military.* New York: New York University Press, 1998.

Higate, Paul. *Military Masculinities: Identity and the State.* Westport, CT: Praeger, 2003.

Higonnet, Margaret, Jane Jenson, Sonya Michel, and Margaret Collins Weitz. *Behind the Lines: Gender and the Two World Wars.* New Haven, CT: Yale University Press, 1987.

Hoffman, John. 2001. *Gender and Sovereignty: Feminism, the State, and International Relations.* New York: Palgrave.

Holm, Jeanne. *Women in the Military: Unfinished Revolution.* Novato, CA: Presidio Press, 1982.

Hooper, Charolette. *Manly States: Masculinities, International Relations, and Gender Politics.* New York: Columbia University Press, 2001.

Horton, Frank B. III, Anthony C. Rogerson and Edward L. Warner III. *Comparative Defense Policy.* Baltimore, MD: Johns Hopkins University Press, 1974.

Howard III, John W. and Laura C. Prividera. "Rescuing Patriarchy or Saving 'Jessica Lynch': The Rhetorical Construction of the American Woman Soldier." *Women and Language.* Vol. 27, no. 2 (2004): 89–97.

Howes, Ruth, and Michael Stevenson. "The Impact of Women's Use of Military Force." In Ruth Howes and Michael Stevenson (eds.), *Women and the Use of Military Force*, 207–217. Boulder, CO: Lynne Rienner, 1993.

Htun, Mala. *Culture, Institution, and Gender Inequality in Latin America.* New York: Basic Books, 2000.

Htun, Mala N. "Mujeres y poder político en Latinoamérica." *Mujeres en el Parlamento. Más allá de los numerous.* Stockholm, Sweden: International IDEA, 2002.

Htun, Mala and Mark Jones. "Engendering the Right to Participate in Decision-making: Electoral Quotas and Women's Leadership in Latin America." In Nikki Craske and Maxine Molyneux (eds.), *Gender and the Politics of Rights and Democracy in Latin America*, 32–56. New York: Palgrave Macmillan, 2002.

Hubert, Don. "The Landmine Treaty: A Case Study in Humanitarian Advocacy." Watson Institute for International Studies, Occasional Paper #42, 2000.

Hudson, Heidi. "Mainstreaming Gender in Peacekeeping Operations: Can Africa Learn from International Experience?" *African Security Review*. Vol. 9, no. 4 (2000): 18–33.

Hudson, Heidi. "'Doing' Security as though Humans Matter: A Feminist Perspective on Gender and the Politics of Human Security." *Security Dialogue*. Vol. 36, no. 2 (June 2005): 155–174.

Hudson, Natalie Florea. *Gender, Human Security and the UN: Security Language* as a Political Framework for Women. London: Routledge, 2009.

Hunter, William, Jr. "Military Manpower Alternatives for the All Volunteer Force in the 1980s/1990s." Carlisle Barracks, PA: Army War College. April 15, 1985. Available at: http://stinet.dtic.mil/cgi-bin/GetTRDoc?AD=ADA157202&Location=U2&doc=GetTRDoc.pdf (accessed 31 July, 2008).

Huntington, Samuel. *The Soldier and the State: The Theory and Politics of Civil-Military Relations*. Cambridge, MA: Belknap Press, 1957.

Huysmans, Jef. "Security! What Do You Mean?: From Concept to Thick Signifier." *European Journal of International Relations*. Vol. 4, no. 2 (1998): 226–255.

Huysmans, Jef. "The European Union and the Securitization of Migration." *JCMS: Journal of Common Market Studies*. Vol. 38 (2000): 751–777.

Ibrahim, M. "Securitization of Migration: A Racial Discourse 1." *International Migration*. Vol. 43, (2005): 163–187.

Inglehart, Ronald and Pippa Norris. "The True Clash of Civilization." *Foreign Policy*. Vol. 135, (2003a): 62–70.

Inglehart, Ronald and Pippa Norris. *Rising Tide: Gender Equality and Cultural Change Around the World*. Cambridge: Cambridge University Press, 2003b.

International Institute for Strategic Studies (IISS). *Military Balance 2006–2007*. London: Routledge/IISS, 2007.

Irigaray, L. *Speculum of the Other Woman*, trans. G.C. Gill. New York: Cornell University Press, 1985.

Isaksson, Eva (ed.). *Women and the Military System*. New York: Harvester Wheatsheaf, 1988.

Iskra, Darlene, Marcia Leithauser, Stephen Trainor, and Mady Segal. "Women's Participation in Armed Forces Cross-Nationally: Expanding Segal's Model." *Current Sociology*. Vol. 50, no. 4 (2002): 771–797.

Jancar-Webster, Barbara. *Women in the Yugoslav National Liberation Movement*. Philadelphia, PA: University of Pennsylvania, 1999.

Janowitz, Morris, and Charles Moskos. "Five Years of the All-Volunteer Force: 1973–1978." *Armed Forces and Society*. Vol. 5 (1979): 171–218.

Jensen, Kimberly. "A Base Hospital is Not a Coney Island Dance Hall: American Woman Nurses, Hostile Work Environment, and Military Rank in the First World War Frontiers." *A Journal of Women Studies*. Vol. 26, no. 2 (2005): 206–235.

Jones, David E. *Women Warriors: A History*. Washington, DC: Brassey's, 1997.

Jones, Kathleen. Dividing the Ranks: Women and the Draft. *Women & Politics*. Vol. 4, no. 4 (1984): 75–87.

Jung, Nora, "Feminist Discourse on Central and Eastern Europe: Hungarian Women's Groups in the Early 1990s as a Case Study." In Alena Heitlinger (ed.), *Émigré Feminism: Transnational Perspectives*, 95–114, Toronto, ON: University of Toronto Press, 1996.

Kaldor, Mary. *New and Old Wars: Organized Violence in a Global Era.* Palo Alto, CA: Stanford University Press, 1999.

Kaldor, Mary. *Elaborating the "New War" Thesis.* New York: Frank Cass, 2005.Kampwirth, Karen. *Women & Guerrilla Movements.* University Park, PA: Pennsylvania State University Press, 2002.

Kaplan, Gisela. *Contemporary Western European Feminism.* New York: New York University Press, 1992.

Karam, Azza. "Women in War and Peacebuilding: The Roads Traversed, the Challenges Ahead." *International Feminist Journal of Politics.* Vol. 3, no. 1 (2001): 2–25.

Karame, Kari. "Military Women in Peace Operations: Experiences of the Norwegian Battalion in UNIFIL 1978–1998." In Louise Olsson and Torunn L. Tryggestad (eds.), *Women and International Peacekeeping*, 85–96. London: Frank Cass, 2001.

Karpinski, Janis. *One Women's Army: The Commanding General of Abu Ghraib Tells Her Story.* New York: Miramax Books, 2006.

Karvonen, Lauri, and Per Selle. *Women in Nordic Politics: Closing the Gap.* Dartmouth, NH: Aldershot, 1995.

Katzenstein, Mary Fainsod. *Faithful and Fearless: Moving Feminist Protest inside the Church and Military.* Princeton, NJ: Princeton University Press, 1998.

Keck, Margaret E. and Kathryn Sikkink. *Activists beyond Borders: Advocacy Networks in International Politics.* Ithaca, NY: Cornell University Press, 1998.

Kennedy-Pipe, Carolyn, and Stephen Welch. "Women in the Military: Future Prospects and Ways Ahead." In Alex Alexandrou, Richard Bartle, and Richard Holmes (eds.), *New People Strategies for the British Armed Forces*, 46–69. Basingstoke: Palgrave Macmillan, 2002. Keohane, Robert. *After Hegemony: Cooperation and Discord in the World Political Economy.* Princeton, NJ: Princeton University Press, 1984.

Keohane, Robert. *Neorealism and Its Critics.* New York: Columbia University Press, 1986.

Keohane, Robert, "Beyond Dichotomy: Conversations between International Relations and Feminist Theory." *International Studies Quarterly.* Vol. 42, no. 1 (1998): 193–198.

Keohane, Robert, and Joseph Nye. *Power and Interdependence.* Glenview, IL: Scott, Foresman & Co., 1989.

Khanna, Ranjana. "Reflections on Sara Ruddick's 'Maternal Thinking'." *Women's Studies Quarterly.* Vol. 37, no. 3/4 (2009): 302–304.

King, Gary, and Christopher Murray. "Rethinking Human Security." *Political Science Quarterly.* Vol. 116, no. 4 (2002): 206–235.

Kirk, Tim and Anthony McElligott. *Opposing Fascism: Community, Authority and Resistance in Europe*. Cambridge: Cambridge University Press, 2004.

Klick, Jean. "Utilization of Women in the NATO Alliance." *Armed Forces and Society*. Vol. 4 (1978): 673–678.

Kriger, N.J. *Guerrilla Veterans in Post-war Zimbabwe: Symbolic and Violent Politics, 1980–1987*. New York: Cambridge University Press, 2003.

Kristeva, J. *Desire in Language: A Semiotic Approach to Literature and Art*. Oxford: Blackwell, 1980.

Krizbai, Janos. "Women Soldiers and Human Resources Policy." *Minerva: Quarterly Report on Women and the Military*. Vol. 17 (1999): 85–91.

Krwawicz, M. Ney. *Women Soldiers of the Polish Home Army*. London: London Branch of the Polish Home Army Ex-Servicemen Association, 2002. Available at: 2.http://www.polishresistance ak.org/PR_WWII_texts_En/12_Article_ En.pdf (accessed June 20, 2009).

Kulish, Nicholas. "As Draft Ends, Polish Military Faces Struggle to Modernize." *New York Times*, December 11, 2008.

Kümmel, Gerhard. "Complete Access: Women in the Bundeswehr and Male Ambivalence." *Armed Forces & Society*. Vol. 28 (2002a): 555–573.

Kümmel, Gerhard. "When Boy Meets Girl: The 'Feminization' of the Military." *Current Sociology*. Vol. 50, no. 5 (2002b): 615–639.

Kwasniewski, Alexander. "Poland and NATO." In George A. Joulwan and Roger Weissinger-Baylon (eds.), *European Security: Beginning a New Century*. XIIIth NATO Workshop, On Political-Military Decision-Making, Warsaw, Poland, June 19–23, 1996. Supreme Allied Center for Strategic Decision Research, 1996. Available at: http://www.csdr.org/Rosati.htm (accessed July 2, 2009).

Lancaster, John. "Aspin to Open Combat Roles to Women: Pentagon is Ready to Lift Ban of Female Pilots in Attack Aircraft." *Washington Post*, April 28, 1993, A1.

Lantis, Jeffrey S. "Strategic Culture and National Security Policy." *International Studies Review*. Vol. 4, no. 3 (2002): 87–113.

Latella, Maria. "Cocktail e politica, le 'donne di Fini' cercano spazio." *Corriere Della Sera*, June 1, 2002, 9.

Lazreg, Marnia. *The Eloquence of Silence: Algerian Women in Question*. New York: Routledge, 1994.

Leen, M. *The European Union, HIV/AIDS and Human Security*. Dublin: Dochas, 2004.

Leonard, Elizabeth D. *Yankee Women: Gender Battles in the Civil War*. New York: W.W. Norton & Co., 1994.

Leonard, Elizabeth D. *All the Daring of the Soldier: Women of the Civil War Armies*. New York: W.W. Norton & Co., 1999.

Levin, Martin. *Feminism and Freedom*. Edison, NJ: Transaction Books, 1987.

Lindley-French, Julian, and Franco Algieri. *A European Defense Strategy*. Gutersloh, Germany: Bertelsmann Foundation, 2004.

Lindsay, James E. *Daily Life in the Medieval Islamic World*. Westport, CT: Greenwood Publishing Group, 2005.

Ling, L.H.M. "Feminist IR: From Critique to Reconstruction." *Journal of International Communication*. Vol. 3, no. 1 (1996): 26–41.

Ling, L.H.M. "Hypermasculinity on the Rise, Again: A Response to Fukuyama on Women and World Politics." *International Feminist Journal of Politics*. Vol. 2, no. 2 (2000): 278–286.

Lipari, Rachel N., and Anita R. Lancaster. *Armed Forces 2002 Sexual Harassment Survey*. Arlington, VA: Defense Manpower Data Center, November 2003. Available at: http://www.defenselink.mil/news/Feb2004/d20040227shs1.pdf.

Locher, Birgit, and Elisabeth Prügl. "Feminism and Constructivism: Worlds Apart or Sharing the Middle Ground?" *International Studies Quarterly*. Vol. 45, no. 1 (2001a): 111–129.

Locher, Birgit and Elisabeth Prügl. "Constructivism's Other Pedigree." In Karin M. Fierke and Knud Erik Jorgensen (eds.), *Constructing International Relations: The Next Generation*, 76–92. Armonk: M.E. Sharpe, 2001b.

Longo, Patrizia. "Italian Feminisms: Past, Present, Future." Paper prepared for delivery at the Annual Meeting of the American Political Science Association, Philadelphia, PA, August 28–31, 2003.

Lorentzen, Lois Ann, and Jennifer Turpin. *The Women and War Reader*. New York: New York University Press, 1988.

Lovenduski, Joni. "Introduction: The Dynamics of Gender and Party." In Joni Lovenduski and Pippa Norris (eds.), *Gender and Party Politics*, 1–15. London: Sage Publications, *1993*.

Luintel, Youba Raj. "Do Males Always Like War? A Critique on Francis Fukuyama and His Hyper Masculine Assertions on 'Feminization of World Politics.'" *Occasional Papers in Sociology and Anthropology*. Vol. 9 (2005): 278–290.

Lundberg, Norma. "Making Sense of War: Demythologizing the Male Warrior." *Atlantis: A Women's Studies Journal*. Vol. 12 (1986): 97–110.

Lyons, Tanya. *Guns and Guerilla Girls: Women in the Zimbabwean Liberation Struggle*. Trenton, NJ: Africa World Press. 2004.

McDevitt, Theresa. *Women and the American Civil War: An Annotated Bibliography*. Westport, CN: Praeger, 2003.

McDonald, R.A. *Women in Combat: When the Best Man for the Job is a Woman*. Maxwell Air Force Base, AL: Air War College, 1991.

McGann, D.R. *Eliminating the Combat Exclusion: Solution to a 25-Year Old Problem*. Carlisle Barracks, PA: Army War College, 1998.

MacKenzie, Megan. "Securitization and Desecuritization: Female Soldiers and the Reconstruction of Women in Post-Conflict Sierra Leone." *Security Studies*. Vol. 18, no. 2 (2009): 241–261.

Mack, A. *Human Security Report: War and Peace in the 21st Century*. New York: Oxford University, 2005.

Mahnken, Thomas G. *United States Strategic Culture*. Report for Defense Threat Reduction Agency Advanced Systems and Concepts Office. 2006.

Available at: http://www.au.af.mil/au/awc/awcgate/dtra/mahnken_strat_ culture.pdf (accessed June 2012).

Malagreca, Miguel, "Lottiamo Ancora: Reviewing One Hundred and Fifty Years of Italian Feminism." *Journal of International Women's Studies*. Vol. 7, no. 4 (May 2006): 69–89.

Mankayi, Nyameka "Male Constructions and Resistance to Women in the Military." *Scientia Militaria: South African Journal of Military Studies*. Vol. 34, no. 2 (2006): 44–64.

Mann-Wall, Barbra. "Sisters of the Holy Cross in the Spanish-American War." *Nursing History Review: Official Journal of the American Association for the History of Nursing*. Vol. 9 (2001): 55–78.

Marchand, Marianne. "The Future of Gender and Development after 9/11: Insights from Postcolonial Feminism and Transnationalism." *Third World Quarterly*. Vol. 30, no. 5 (2009): 921–935.

Marchand, Marianne H. and Anne Sisson Runyan. "Introduction: Feminist Sightings of Global Restructuring: Conceptualizations and Reconceptualizations." In H. Marianne Marchand and Anne Sisson (eds.), *Gender and Global Restructuring: Sightings, Sites and Resistance*, 1–22. London: Routledge, 2000.

Marhia, Natasha. "some humans are more human than others: Troubling the 'Human' in Human Security from a Critical Feminist Perspective." *Security dialogue*. Vol. 44, no.2 (2013): 19–35.

Marmion, Harry A. *The Case against a Volunteer Army*. Chicago, IL: Quadrangle Books, 1971.

Martino, Wayne. "'Cool Boys', 'Party Animals', 'Squids' and 'Poofters': Interrogating the Dynamics and Politics of Adolescent Masculinities in School", *The British Journal of the Sociology of Education*. Vol. 20, no. 2 (1999): 239–263.

Martinusz, Zoltan. *Defense Reform in Hungary: A Decade of Strenuous Efforts and Missed Opportunities*. Dulles, VA: Brassey's, 2002.

Mazur, Amy. *Theorizing Feminist Policy*. Oxford: Oxford University Press, 2002.

Mazur, Amy. "The Impact of Women's Participation and Leadership on Policy Outcomes: A Focus on Women's Policy Machineries." Paper prepared for the United Nations Expert Group Meeting on Equal Participation of Women and Men in Decision-Making Processes, October 24–27, 2005. Available at: http://www.un.org/womenwatch/daw/egm/eql-men/docs/EP.5_Mazur.pdf (accessed March 2009).

Mazurana, Dyan, Angela Raven-Roberts, and Jane Parpart. *Gender, Conflict, and Peacekeeping*. Lanham, MD: Rowman and Littlefield, 2005.

Mearsheimer, John. "The False Promise of International Institutions." *International Security*. Vol. 19, no. 3 (1995): 5–49.

Meilinger, Philip. "Busting the Icon: Restoring Balance to the Influence of Clausewitz." *Strategic Studies Quarterly*. Vol. 1, no. 1 (2007):116–145.

Meriwether Wingfield, Nancy and Maria Bucur. *Gender and War in Twentieth Century Eastern Europe*. Bloomington, IN: University of Indiana Press, 2006.

Merryman, Molly. *Clipped Wings: The Rise and Fall of the Women Airforce Service Pilots (WASPS) of World War II*. New York: New York University Press, 1998.

Meyer, Leisa D. *Creating GI Jane: Sexuality and Power in the Women's Army Corps during World War II*. New York: Columbia University Press, 1996.

Meyer, Mary K. and Elisabeth Prügl. *Gender Politics in Global Governance*. Lanham, MD: Rowman and Littlefield Publishers, 1999.

Mileham, Patrick. "Amateurs, Conscripts, Citizens, Professionals: How Do Armed Forces Measure Up?" *Defense and Security Analysis*. Vol. 21, no. 2 (2005): 213–216.

Miller, Laura. "Not Just Weapons of the Weak: Gender Harassment as a Form of Protest for Army Men." *Social Psychology Quarterly*. Vol. 60, no. 1 (1997): 32–35.

Milner, Helen. "The Assumption of Anarchy in International Relations Theory: A Critique." *Review of International Studies*. Vol. 17, no. 1 (1991): 67–85.

Mitchell, Brian. *Women in the Military: Flirting with Disaster*. Washington, DC: Regnery Publishing, 1998.

Mohanty, Chandra, and Ann Russo. *Third World Women and the Politics of Feminism*. Bloomington, IN: University of Indiana Press, 1991.

Monahan, Evelyn, and Rosemary Neidel-Greenlee. *A Few Good Women: America's Military Women from World War I to the Wars in Iraq and Afghanistan*. New York: Random House Digital, 2011.

Moon, Katharine. *Sex among Allies: Militarized Prostitution in U.S.–South Korea Relations*. New York: Columbia University Press, 1997.

Moore, Brenda. "From Underrepresentation to Overrepresentation: African American Women." In Judith Stiehm (ed.), *It's Our Military, Too!: Women and the US Military*, 115–135. Philadelphia, PA: Temple University Press, 1996.

Morden, Betti. *The Women's Army Corp, 1945–1978*. Washington, DC: Center of Military History United States Army, 2000.

Morgan, Matthew J. *The American Military after 9/11: Society, State, and Empire*. New York: Palgrave Macmillan, 2008.

Morgenthau, Hans J. *Politics among Nations: The Struggle for Power and Peace*. 5th edn. New York: Knopf, 1973.

Moser, Caroline, and Fiona Clark. *Victims, Perpetrators or Actors?: Gender, Armed Conflict and Political Violence*. London: Zed Books, 2001.

Moskos, Charles. "National Service and the All-Volunteer Force." *Social Science and Public Policy*. Vol. 17, no. 1 (1979): 70–72.

Moskos, Charles. "Making the All-Volunteer Force Work: A National Service Approach." *Foreign Affairs*. Vol. 60 (1981): 17–34.

Moskos, Charles. "Female GIs in the Field." *Society*. Vol. 22, no. 6 (1985): 28–33.

Moskos, Charles and Frank R. Wood. *The Military: More Than Just a Job?*. McLean, VA: Pergamon-Brassey's International Defense Publishers, 1988.

Moskos, Charles. "Army Women." *Atlantic Monthly*. Vol. 266, no. 2 (1990): 71–78.

Moskos, Charles, John Allen Williams, and David R. Segal. *The Postmodern Military: Armed Forces After the Cold War*. New York: Oxford University Press, 1999.

Mrozik, Agnieszka. "Sexual Harassment in Polish Army: General Dismissed." *Poland National VAW Monitor*, May 28, 2006.

Murawiec, Laurent and David M. Adamson (eds.). *Demography and Security, Proceedings of a Workshop, Paris, France, November 2000*. Santa Monica, CA: RAND, 2000.

Murray, Douglas J. and Paul R. Viotti (eds.). *The Defense Policies of Nations: A Comparative Study*. Baltimore, MD: Johns Hopkins University Press, 1982.

Murray, Douglas and Paul R. Viotti (eds.). *The Defense Policies of Nations a Comparative Study*. 3rd edn. Baltimore, MD: Johns Hopkins University Press, 1994.

Mutimer David Roger. "My Critique is Bigger than Yours: Constituting Exclusions in Critical Security Studies." *Studies in Social Justice*. Vol. 3, no. 1 (2009): 9–20.

Najam, A. *Environment, Development and Human Security: Perspectives from South Asia*. Lanham, MD: University Press of America, 2003.

Nantais, Cynthia and Lee, Martha F. "Women in the United States Military: Protectors or Protected? The Case of Prisoner of War Melissa Rathbun-Nealy." *Journal of Gender Studies*. Vol. 8, no. 2 (1999): 181–192.

Naples, Nancy A. and Manisha Desai (eds.). *Women's Activism and Globalization: Linking Local Struggles and Transnational Politics*. New York: Routledge, 2002.

Narayan, Uma. *Dislocating Cultures: Identities, Traditions, and Third-World Feminism*. New York: Routledge, 1997.

NATO. "Hungary National Report." In *50th Anniversary Women in the NATO Forces Report*, 30–32. Brussels: NATO HQ, Office on Women in the NATO Forces, 1999a. Available at: http://www.nato.int/ims/1999/win/report99.pdf (accessed December 2012).

NATO. "Poland National Report." In *50th Anniversary Women in the NATO Forces Report*, 43–46. Brussels: NATO HQ, Office on Women in the NATO Forces, 1999b. Available at: http://www.nato.int/ims/1999/win/report99.pdf (accessed December 2012).

Newmwan, E. "Human Security and Constructivism." *International Studies Perspectives*. Vol. 2, no. 3 (2001): 239–251.

Newman, Edward. 2010. "Human Security." International Studies Encyclopedia Online.

Nielsen, Vikci. "Women in Uniform." *NATO Review*. Vol. 49 (2001): 30–34.

Nichiporuk, Brian. *The Security Dynamics of Demographic Factors*. Santa Monica, CA: RAND, 2000.

Nordheimer, Jon. "Women's Role in Combat: The War Resumes." *New York Times*, May 26, 1991.

Norland, Ron. "For Soldiers, Death Sees No Gender Lines." *New York Times*, June 11, 2011. Available at: http://www.nytimes.com/2011/06/22/world/asia/22afghanistan.html?pagewanted=all (accessed August 14, 2011).

Norris, Pipa and Joni Lovenduski. "Women Candidates for Parliament: Transforming the Agenda?" *British Journal of Political Science*. Vol. 19, no. 1 (1989): 106–115.

Nuciari, Marina. *Women in the Military: Sociological Arguments for Integration.* New York: Kluwer Academic/Plenum, 2003.

Olsson, Louise and Torunn L. Tryggestad (eds.). *Women and International Peacekeeping*. London: Routledge, 2001.

Owens, Mackubin Thomas. *US Civil-Military Relations after 9/11: Renegotiating the Civil-Military Bargain*. New York: Continuum, 2011.

Palm Center, The. "Gang Members Get Trained in the Army." March 8, 2008. Available at: http://www.palmcenter.org/press/dadt/in_print/gang_members_get_trained_army (accessed November 5, 2012).

Pampell Conaway, C. and J. Shoemaker. *Women in United Nations Peace Operations: Increasing the Leadership Opportunities*. Georgetown University, Washington, DC: Women in International Security, 2008.

Paoletti, Ciro. *A Military History of Italy*. Santa Barbara, CA: Praeger International Security, 2008.

Paris, Roland. "Human Security: Paradigm Shift or Hot Air?" *International Security*. Vol. 26, no. 2 (2001): 87–102.

Parpart, Jane. L. "Exploring the Transformative Potential of Gender Mainstreaming in International Development Institutions." *Journal of International Development*. Online. October, 2013.

Parrish, Karen. "DOD Opens More Jobs, Assignments to Military Women." American Forces Press Service, February 9, 2012. Available at: http://www.defense.gov/news/newsarticle.aspx?id=67130 (accessed November 10, 2012).

Penn, Shana. *Solidarity's Secret: The Women Who Defeated Communism in Poland*. Ann Arbor, MI: University of Michigan Press, 2006.

Peterson, V. Spike. *"Clarification and Contestation: A Conference Report on 'Women, the State and War': What Difference Does Gender Make?"* Los Angeles, CA: Center for International Studies, University of Southern California, 1989.

Peterson, V. Spike. "Whose Rights? A Critique of the 'Givens' in Human Rights Discourse." *Alternatives*, Vol. 15, no. 3 (1990): 303–344.

Peterson, V. Spike (ed.). *Gendered States: Feminist (Re)Visions of International Relations Theory*. Boulder, CO: Lynne Rienner, 1992a.

Peterson, V. Spike. *Security and Sovereign States: What Is at Stake in Taking Feminism Seriously?* Boulder, CO: Lynne Rienner, 1992b.

Peterson, V. Spike and Anne Sisson Runyan. *Global Gender Issues: Dilemmas in World Politics*. Boulder, CO: Westview Press, 1999.

Pettman, Jan Jindy. *Worlding Women*. New York: Routledge, 1996.

Pietilä, Hilkka and Jeanne Vickers. *Making Women Matter: The Role of the United Nations*. London: Zed Books, 1990.

Pin-Fat, Veronique and Maria Stern. "The Scripting of Private Jessica Lynch: Biopolitics, Gender and the 'Feminization' of the U.S. Military." *Alternatives*. Vol. 30, no. 1 (2005): 25–53.

Poggioli, Sylvia. "In Italy, Feminism Out, Sex Symbols In." *NPR Morning Edition*, December 3, 2008. Transcripts of the show available at: http://www.npr.org/templates/story/story.php?storyId=97402636 (accessed June 10, 2009).

Pojmann, Wendy. *Italian Women and International Cold War Politics, 1944–1968*. New York: Fordham University Press, 2013.

Prügl, Elisabeth. *The Global Construction of Gender: Home-Based Work in the Political Economy*. New York: Columbia University Press, 1999.

Quester, Aline, O. and Gilroy, Curtis L. "Women and Minorities in America's Volunteer Army." *Contemporary Economic Policy*. Vol. 20, no. 2 (2002): 111–112.

Raible, Karen. "Compulsory Military Service and Equal Treatment of Men and Women – Recent Decisions of the Federal Constitutional Court and European Court of Justice." *German Law Journal*. Vol. 4, no. 4 (2003): 239.

Razavi, S. and C. Miller. "Gender mainstreaming: A Study of Efforts by the UNDP, the World Bank, and the ILO to Institutionalize Gender Issues." New York: United Nations Research Institute for Social Development Occasional Paper no.4, 1995.

Reanda, Laura. "Engendering the United Nations: The Changing International Agenda." *European Journal of Women's Studies*. Vol. 6 (1999): 49–68.

Reardon, Betty. *Sexism and the War System*. New York: Columbia University Teachers College, 1985.

Reardon, Betty. *Women and Peace: Feminist Visions of Global Security*. Albany, NY: SUNY Press, 1993.

Rehn, Elizabeth and Ellen Johnson. *Women, War, and Peace: The Independent Experts' Assessment of the Impact of Armed Conflict on Women and Women's Role in Peace-Building*. New York: UNIFEM, 2002.

Reif, Linda L. "Women in Latin American Guerrilla Movements: A Comparative Perspective." *Comparative Politics*. Vol. 18, no. 2 (1986): 147–169.

Reynolds, Andrew. "Women in the Legislatures and Executives of the World: Knocking at the Highest Glass Ceiling." *World Politics*. Vol. 51, no. 4 (1999): 547–572.

Riddell-Dixon, Elizabeth. *Canada and Beijing Conference on Women: Governmental Politics and NGO Participation*. Vancouver, CA: University of British Columbia Press, 2001.

Rimalt, Noya. "Women in the Sphere of Masculinity: The Double-Edged Sword of Women's Integration in the Military." *Duke Journal of Gender Law and Policy*. Vol. 14, no. 2 (2007): 1097–2007.

Risen, James. "Military Has Not Solved Problem of Sexual Assault, Women Say." *New York Times*, November 2, 2012, A15.

Risse-Kappen, Thomas, Stephen C. Ropp, and Kathryn Sikkink. *The Power of Human Rights: International Norms and Domestic Change.* Cambridge: Cambridge University Press, 1999.

Rodano, Marisa. *Memorie di una che c'era. Una storia dell'Udi.* Milan: Il Saggiatore, 2010.

Rodriguez, Victoria (ed.). *The Emerging Role of Women in Mexican Political Life.* Boulder, CO: Westview Press, 1998.

Rogers, Selwyn P. Jr. "An All-Volunteer Force." *Military Review.* Vol. 50, no. 9 (1970): 89–95.

Rosati, Dariusz. "New Security Architecture in Europe: A Polish View." In George A. Joulwan and Roger Weissinger-Baylon (eds.), *European Security: Beginning a New Century.* XIIIth NATO Workshop, On Political-Military Decision-Making, Warsaw, Poland, June 19–23, 1996. Supreme Allied Center for Strategic Decision Research, 1996. Available at: http://www.csdr.org/96Book/Rosati.htm (accessed November 2010).

Roosevelt, Eleanor. "My Day." GWU Collection of "My Day" columns. Available at: http://www.gwu.edu/~erpapers/myday/displaydoc.cfm?_y=1942&_f=md056279 (accessed April 2013).

Rostker, Bernard D. *I Want You: The Evolution of the All-Volunteer Force.* Santa Monica, CA: RAND Corporation, 2006.

Ruddick, Sara. "Preservative Love and Military Destruction: Some Reflections on Mothering and Peace." In Joyce Trebilcot (ed.), *Mothering: Essays in Feminist Theory,* 231–262. Totowa, NJ: Rowman & Allanheld, 1983a.

Ruddick, Sara. "Pacifying the Forces: Drafting Women in the Interests of Peace." *Signs: Journal of Women in Culture and Society.* Vol. 8, no. 3 (1983b): 471–489.

Ruddick, Sara. *Maternal Thinking: Toward a Politics of Peace.* Boston, MA: Beacon Press, 1989.

Runyan, A. Sisson. "Gender Relations and the Politics of Protection." *Peace Review.* Special Issue on Women, Men and the State 2. Vol. 2, no. 4, (1990): 28–31.

Runyan, A. Sisson. "Still Not 'At Home' in IR: Feminist World Politics Ten Years Later." *International Politics.* Vol. 39, no. 3 (2002): 361–368.

Runyan, A. Sisson and V. Spike Peterson. "The Radical Future of Realism: Feminist Subversions of IR Theory." *Alternatives.* Vol. 16 (1991): 67–106.

Russell, Richard L. "NATO's European Members: Partners or Dependents?" *Naval War College Review.* Vol. 56, no. 1 (2003): 30–40.

Russet, Bruce. "Reintegrating the Subdisciplines of International and Comparative Politics." *International Studies Review.* Vol. 5, no. 4 (2003): 9–12.

Sagawa, Shirley and Nancy Duff Campbell. *Women in the Military Issue Paper: Women in Combat.* Washington DC: National Women's Law Center, 1992.

Saimons, V.J. *Women in Combat: Are the Risks to Combat Effectiveness Too Great?.* Fort Leavenworth, KS: U.S. Army Command and General Staff College, School of Advanced Military Studies, 1992.

Sandell, Rickard. "Coping with Demography in NATO Europe: Military Recruitment in Times of Population Decline." In Curtis Gilroy and Cindy Williams (eds.), *Service to Country: Personnel Policy and the Transformation of Western Militaries*, 65–98. Cambridge, MA: JFK School of Government, 2006.

Sanprie, Virginia. "Identity Cleft: Analysis of Identity Construction in Media Coverage of the Jessica Lynch Story." *Feminist Media Studies*. Vol. 5, no. 3 (2005): 388–391.

Santi, Marta. "Working towards Equal Opportunities for Women in Employment." Centro di *Studi Economici* Sociali *e Sindacali* (CESOS). April 10, 2007. Available at: http://www.eurofound.europa.eu/eiro/2006/12/articles/it0612059i.htm (accessed June 5, 2009).

Sartori, Giovanni. "Concepts Misformation in Comparative Politics." *Annual Political Science Review*. Vol. 64 (1970): 1033–1053.

Sasson-Levy, Orna. "Feminism and Military Gender Practices: Israeli Women Soldiers in 'Masculine' Roles." *Sociological Inquiry*. Vol. 73, no. 3 (2003): 440—465.

Saywell, Shelley. *Women in War: From World War II to El Salvador*. Toronto, ON: Penguin Books, 1986.

Schlafly, Phylllis. *Feminist Fantasies*. Dallas, TX: Spence Publishing Company, 2003.

Schmitt, Erick. "Ban on Women in Combat Divides Four Service Chiefs." *New York Times*, June 19, 1991, A16.

Seelye, Katharine Q. "Gingrich's 'Piggies' Poked." *New York Times*, January 19, 1995.

Segal, David R. "How Equal is 'Equity'?" *Society*. Vol.18 (1981): 31–33.

Segal, Mady Wechsler. "The Argument for Female Combatants." In Nancy Loring Goldman (ed.), *Female Soldiers: Combatants or Noncombatants?: Historical and Contemporary Perspectives,* 262–290. Westport, CN: Greenwood Press, 1982.

Segal, Mady Wechsler. "Women's Military Roles Cross-Nationally: Past, Present, and Future." *Gender and Society*. Vol. 9, no. 6 (1995): 757–775. Seifert, Ruth. *Gender and the Military: An Outline of Theoretical Debates*. New Brunswick, NJ: Transaction, 2003.

"Sexual Assault and Violence against Women in the Military and at the Academies." Hearing before the Subcommittee on National Security, Emerging Threats, and International Relations of the Committee on Government Reform, House of Representatives, One Hundred Ninth Congress, second session, June 27, 2006. Vol. 4. Washington: U.S. Government Printing Office, 2007.

Shadrock, Sherri L. *Women in the US Army: A Quiet Revolution in Military Affairs*. Fort Leavenworth, KS: School of Advanced Military Studies United States Army Command and General Staff College, 2007.

Shahidin, Hammed. "Women and Clandestine Politics in Iran, 1970–1985." *Feminist Studies*. Vol. 23, no. 1 (1997): 7–43.

Shanker, Thom and Nicholas Kulish. "Russia Lashes Out on Missile Deal." *New York Times*, August 15, 2008. Available at: http://www.nytimes.com/2008/08/15/world/europe/16poland.html?pagewanted=all&_r=0 (accessed May 2013).

Sherman, Jannan. "'They Either Need These Women or They Do Not': Margaret Chase Smith and the Fight for Regular Status for Women in the Military." *The Journal of Military History*. Vol. 54, no. 1 (1990): 47–78.

Sherman, Janann. *No Place for Woman: A Life of Senator Margaret Chase Smith*. New Brunswick, NJ: Rutgers University Press, 2001.

Sherr, James. "A Corner Turned?" In Natalie Mychajlyszyn and Harald Von Riekhoff (eds.), *The Evolution of Civil–Military Relations in East-Central Europe and the Former Soviet Union*, 63–82. Westport, CT: Greenwood Publishing Group, 2004.

Shields, Patricia M. "Sex Roles in the Military." In Charles Moskos and Frank Wood (eds.), *The Military – More Than a Job?*, 99–114. McLean, VA: Pergamon-Brassey's International Defense Publishers, 1988.

Sills, D.L. (ed.). *International Encyclopedia of Social Science*. Vol. 10. New York: Macmillan and The Free Press, 1972.

Singer, Peter. *Corporate Warriors: The Rise of the Privatized Military Industry.* Ithaca, NY: Cornell University Press, 2003.

Sjoberg, Laura. *Gender, Justice, and the Wars in Iraq: A Feminist Reformulation of Just War Theory*. Lanham, MD: Lexington Books, 2006.

Sjoberg, Laura. "Looking Forward, Conceptualizing Feminist Security Studies." *Politics & Gender*. Vol. 7, no. 4 (2011): 600–604.

Sjoberg, Laura and Sandra Via. *Gender, War, and Militarism: Feminist Perspectives*. Santa Barbara, CA: ABC-CLIO, 2010.

Sjoberg, Laura and Jillian Martin. "Feminist Security Theorizing." In Bob Denemark, (ed.), *International Studies Compendium.*, 2007. Available at: http://sitemason.vanderbilt.edu/files/ieKk5G/Laura%20Sjoberg%20Jillian%20Martin%20Compendium%20Contribution.pdf (accessed January 2013).

Sloan, Elinor Camille. *The Revolution in Military Affairs: Implications for Canada and NATO*. Montreal, ON: McGill-Queen's University Press, 2002.

Smith, D.W. and Mowery, D.L. *Women in Combat: What Next? (Final Report)*. Newport, RI: Naval War College, 1992.

Snitow, Ann. "The Church Wins, Women Lose: Poland's Abortion Law." *The Nation*, vol. 256, April 26, 1993. Available at: http://www.thenation.com/archive/detail/9304130117 (accessed July 1, 2009).

Snow, Donald. *National Security in a New Era*. New York: Pearson, 2007.

Snyder, Jack. *Domestic Politics and International Ambition*. Ithaca, NY: Cornell University Press, 1991.

Soderbergh, Peter A. *Women Marines in the Korean War Era*. Westport, CT: Praeger, 1994.

Solaro, Erin. *Women in the Line of Fire*. Emeryville, CA: Seal Press, 2006.

Solaro, Erin. "Introductions: An Unabashed Feminist Writes about Women in the Military." February 17, 2010. Available at: http://www.pbs.org/pov/

regardingwar/conversations/women-and-war/introductions-an-unabashed-feminist-writes-about-women-in-the-military.php (accessed June 2012).

Solheim, Bruce O. *On Top of the World: Women's Political Leadership in Scandinavia and Beyond.* Westport, CT: Greenwood Press, 2000.

Sorokin, Ellen. "Patriotism Touted as Lure for Recruits." *The Washington Times,* August 17, 2002.

Sowers, Susan R. *Women Combatants in World War I: A Russian Case Study.* Carlisle Barracks, PA: U.S. Army War College, 2003. Available at: http://handle.dtic.mil/100.2/ADA414547 (accessed March 20, 2013).

Spivak, Gayatri Chakravorty. *A Critique of Postcolonial Reason: Toward a History of the Vanishing Present.* Cambridge, MA: Harvard University Press, 1999.

Steans, Jill. *Gender and International Relations: An Introduction.* New Brunswick, NJ: Rutgers University Press, 1998.

Sterling-Folker, Jennifer. *Making Sense of International Relations (IR) Theory.* Boulder, CO: Lynne Rienner, 2006.

Stevens, Gwendolyn and Sheldon Gardner. "But Can She Command a Ship? Acceptance of Women by Peers at the Coast Guard Academy." *Sex Roles.* Vol. 16, no. 3–4 (1987): 181–188.

Stiehm, Judith Hicks. *Bring Me Men and Women: Mandated Change at the U.S. Air Force Academy.* Berkeley, CA: University of California Press, 1981.

Stiehm, Judith Hicks. "Arms and the Enlisted Women. Philadelphia, PA: Temple University Press, 1989.

Stiehm, Judith Hicks (ed.). *It's Our Military, Too: Women and the U.S. Military.* Philadelphia, PA: Temple University Press, 1996.

Stiehm, Judith Hicks. "Women, Peacekeeping and Peacemaking: Gender Balance and Mainstreaming." *International Peacekeeping.* Vol. 8, no. 2 (2001): 39–48.

Stoddard, E.R. "Female Participation in the U.S. Military. Gender Trends by Branch, Rank and Racial Categories." *Minerva: Quarterly Report on Women and the Military.* Vol. 11, no.1 (1993): 23–35.

Studlar, Donley T. and Ian McAllister. "Does a Critical Mass Exist? A Comparative Analysis of Women's Legislative Representation since 1950." *English Journal of Political Research.* Vol. 41, no. 2 (2002): 233–253.

Sunga, Lyal. *The Concept of Human Security: Does It Add Anything of Value to International Legal Theory or Practice?* Farnham: Ashgate, 2009.

Swers, Michele L. *The Difference Women Make: The Policy Impact of Women in Congress.* Chicago, IL: University of Chicago Press. 2002.

Sylvester, Christine. "Patriarchy, Peace and Women Warriors." In Linda Rennie Forcey (ed.), *Peace: Meanings, Politics, Strategies.* New York: Praeger/Greenwood, 1989.

Sylvester, Christine. *Feminist Theory and International Relations in a Postmodern Era.* New York: Cambridge University Press, 1994.

Sylvester, Christine. "The Contributions of Feminist Theory to International Relations," in Steve Smith, Ken Booth, and Marysia Zalewski (eds.),

International Relations Theory: Positivism and Beyond. New York: Cambridge University Press, 1996.

Sylvester, Christine. *Feminist International Relations: An Unfinished Journey.* New York: Cambridge University Press, 2002.

Sylvester, Christine. "Anatomy of a Footnote." *Security Dialogue.* Vol. 38, no. 4 (2007): 547–558.

Szabo, Andrea. "The Establishment of the Hungarian Women Soldier's Section and Its Justification." *Minerva: Quarterly Report on Women and the Military.* Vol. 17, no. 3–4 (1999): 68–73.

Tacitus. *The Annals of Imperial Rome.* London: Harmondsworth, 1956.

Tanner, Doris Brinker. 1996. *Zoot Suits and Parachutes, and Wings of Silver, Too! The World War II Air Force Training of Women Pilots.* Paducah, KY: Turner Publishing Company.

Taraki, Lisa. "The Role of Women." In Deborah J. Gerner and Jillian Schwedler (eds.), *Understanding the Contemporary Middle East.* 3rd edn., 345–372. Boulder, CO: Lynne Rienner, 2008.

Tashakkori, Abbas and Charles Teddlie (eds.). *Handbook of Mixed Methods in Social and Behavioral Research.* Thousand Oaks, CA: Sage Publications, 2003.

Taylor, Sandra, C. *Vietnamese Women at War: Fighting for Ho Chi Minh and the Revolution.* Lawrence, KS: University Press of Kansas, 1999.

Tendrich Frank, Lisa (ed.). *Women in the American Civil War.* Santa Barbara, CA: ABC-CLIO, 2008.

Thomas, Caroline. 2000. *Global Governance, Development and Human Security the Challenge of Poverty and Inequality.* London and Sterling, VA: Pluto Press.

Thomas, Patricia J. and Zanette Uriells. "Pregnancy and Single Parenthood: Results of a 1997 Survey." San Diego, CA: Naval Personnel Research & Development Center, TR-98–96, September 1998.

Thomas, Patricia J., Carol E. Newel, and Dawn M. Eliassen. "Sexual Harassment of Navy Personnel: Results of a 1993 Survey." San Diego, CA: Naval Personnel Research & Development Center, November 1995.

Thomas, Sue and Susan Welch. "The Impact of Gender on Activities and Priorities of State Legislators." *The Western Political Quarterly*, Vol. 44, No. 2 (1991): 445–456.

Tickner, J. Ann. *Gender in International Relations: Feminist Perspectives on Achieving Global Security.* New York: Columbia University Press, 1992.

Tickner, J. Ann. "You Just Don't Understand: Troubled Engagements between Feminists and IR Theorists." *International Studies Quarterly.* Vol. 41, no. 4 (1997): 611–632.

Tilghman, Andrew and Lance M. Bacon. "DoD to Open 14,000 Jobs to Women." *Army Times*, February 18, 2012. Available at: http://www.armytimes.com/news/2012/02/army-dod-to-open-14000-army-jobs-women-021812w.

Titkov, Anna. *Women in the Politics of Poland.* In Marilyn Rueschemeyer (ed.), *Women in the Politics of Post-Communist Eastern Europe*, 24–32. Armonk, NY: M.E. Sharpe, 1998.

Tuten, Jeff M. "The Argument against Female Combatants." In Nancy Loring Goldman (ed.), *Female Soldiers: Combatants or Noncombatants?: Historical and Contemporary Perspectives*, 237–265. Westport, CT: Greenwood Press, 1982.

Titunic, Regina F. "The First Wave: Gender Integration and Military Culture." *Armed Forces and Society*. Vol. 26, no. 2 (2000): 229–257.

Treadwell, Mattie E. *United States Army in World War II: The Women's Army Corps*. Washington, DC: Defense Department, Army, Center of Military History, 1954.

U.S. Army. *Women in Combat*. (Final Report). Fort Monroe, VA: Training and Doctrine Command, 1986. Available at: http://www.dtic.mil/cgi-bin/GetTRDoc?AD=ADA393292 (accessed October 2012).

U.S. Congress, House, Armed Services Subcommittee no. 3, Hearings on S 1641, 80th Congress, 2nd session, *Congressional Record*, 18 Feb 1948.

U.S. Department of Commerce, Bureau of the Census. "Women in Poland." Population Division, International Programs Center, July 1995. Available at: http://www.census.gov/ipc/prod/women_po.pdf (accessed July 1, 2009).

U.S. Department of Defense, Office of the Secretary of Defense. *Memorandum: Direct Ground Combat Definition and Assignment Rule*. January 13, 1994.

U.S. Department of Defense. *Task Force Report on Care for Victims of Sexual Assault*. Washington, DC: Department of Defense, April 2004. Available at: http://www.defense.gov/news/may2004/d20040513satfreport.pdf.

U.S. Department of State, Under Secretary for Public Diplomacy and Public Affairs Bureau of Public Affairs. "Background Notes, Poland." Bureau of Public Affairs: Electronic Information and Publications Office, January 2009. Available at: http://www.state.gov/r/pa/ei/bgn/2875.htm (accessed July 2, 2009).

U.S. General Accounting Office. "DOD Service Academies: More Actions Needed to Eliminate Sexual Harassment." GAO/NSIAD, 94–96, January 1994. Available at: http://www.gao.gov/assets/160/154095.pdf (accessed March 17, 2010).

U.S. General Accounting Office. *Gender Issues Information on DoD's Assignment Policy and Direct Ground Combat Definition*. Washington, DC: U.S. Government Printing Office, 1998.

U.S. General Accounting Office: *Gender Issues: Changes Would Be Needed to Expand Selective Service Regulations to Women.* GAO/NSAID, 98–99, June 1998. http://www.gao.gov/assets/230/225927.pdf (accessed March 12, 2010).

U.S. General Accounting Office. *Gender Issues: Analysis of Promotion and Career Opportunities Data.* GAO/NSIAD, 98–157. Washington, DC: 1998. Available at: http://www.gao.gov/archive/1998/ns98157.pdf (accessed March 17, 2010).

U.S. General Accounting Office. "*Military Housing: Costs of Separate Barracks for Male and Female Recruits in Basic Training.*" GAO/NSIAD, 99–75, March 1999. Available at: http://www.gao.gov/assets/230/226900.pdf (accessed March 12, 2010).

U.S. House Document no. 108–223, *Women in Congress 1917—2006*. Washington DC: U.S. Government Printing Office, 2006. Available at: http://www.gpoaccess. gov/serialset/cdocuments/hd108–223/part1-chap2.pdf (accessed May 22, 2009).

United Nations. Peacekeeping Factsheet February 2008. Available at: http:// www.un.org/Depts/dpko/factsheet.pdf (accessed May 2, 2009).

United Nations Commission on Human Security. *Human Security Now*. New York: UN, 2003.

Urbelis, Vaidotas. "Impact of NATO Membership on Military Service in the Baltic States." In Curtis Gilroy and Cindy Williams (eds.), *Service to Country: Personnel Policy and the Transformation of Western Militaries*, 97–119. Cambridge, MA: JFK School of Government, 2006.

Van Crevald, Martin. *Men, Women & War: Do Women Belong in the Front Line?* London: Cassell and Co, 2001.

Vickers, Jill. "What Makes Some Democracies More 'Women-Friendly'?." Paper prepared for Annual Meeting of the Canadian Political Science Association, York University, Toronto, June 2006. Available at: http://www.cpsa-acsp.ca/ papers-2006/Vickers.pdf (accessed April 2009).

Villani, Domenico. "Recruitment in a Period of Transformation: The Italian Experience." In C. Gilroy and C. Williams (eds.), *Service to Country: Personnel Policy and the Transformation of Western Militaries*, 381–396. Cambridge, MA: Harvard University Press, 2005.

Viterna, Jocelyn and Kathleen M. Fallon. "Democratization, Women's Movements, and Gender-Equitable States: A Framework for Comparison." *American Sociological Review*. Vol. 73, no.4 (2008): 668–689.

Vuic, Kara Dixon. *Officer, Nurse, Woman: The Army Nurse Corps in the Vietnam War*. Baltimore, MD: Johns Hopkins University Press, 2010.

Wadding, Linda. "Soldier Rape: Don't Ask (for help), Don't Tell (a soul)." *Iowa Independent*, August 6, 2008. Available at: http://iowaindependent.com/3494/ soldier-rape-dont-ask-for-help-dont-tell-a-soul.

Wæver, Ole. "Securitization and Desecuritization." In Ronnie D. Lipschutz (ed.), *On Security*, 46–86, New York: Columbia University Press, 1995.

Wæver, Ole, Barry Buzan, Morten Kelstrup and Pierre Lemaitre. *Identity, Migration and the New Security Agenda in Europe*. London: Pinter, 1993.

Walker, Diana Barnato. *Spreading My Wings: One of Britain's Top Women Pilots Tells Her Remarkable Story*. Sparkford, UK: Patrick Stephens, 1994.

Wallis, Emma. "Why are Italy's Women out of Work?" BBC News Rome, April 9, 2008. Available at: http://news.bbc.co.uk/go/pr/fr/-/2/hi/business/7337145.stm (accessed June 4, 2009).

Waltz, Kenneth. *Theory of International Politics*. New York: McGraw-Hill, 1979.

Warner, John T., Curtis J. Simon, and Deborah M. Payne. *Enlistment Supply in the 1990s: A Study of the Navy College Fund and Other Enlistment Incentive Programs*. Arlington, VA: Defense Manpower Data Center, 2001.

Weber, C. "Good Girls, Little Girls and Bad Girls: Male Paranoia in Robert Keohane's Critique of Feminist International Relations." *Millennium: Journal of International Studies*. Vol. 23, no. 2 (1995): 337–349.

Weedon, C. *Feminist Practice and Poststructuralist Theory*. New York: Blackwell, 1987.

Weinstein, Laurie Lee, and Christie C. White. *Wives and Warriors: Women and the Military in the United States and Canada*. Westport, CT: Bergin & Garvey, 2007.

Westwood, J. and Turner, H. "Marriage and Children as Impediments to Career Progression of Active Duty Career Women Army Officers." Research Report. Carlisle Barracks, PA: Army War College, 1996.

Whelan, Imelda. *Modern Feminist Thought*. New York: New York University Press, 1995.

Whitworth, Sandra. "Gender in the Inter-Paradigm Debate." *Millennium*. Vol. 18, no. 2 (1989): 265–292.

Whitworth, Sandra. "Gender International Relations and the Case of ILO." *Review of International Studies*. Vol. 20, no. 4 (1994a): 389–405.

Whitworth, Sandra. 1994. "Feminist Theories: From Women to Gender and World Politics." In Peter R. Beckman and Francine D'Amico (eds.), *Women, Gender, and World Politics: Perspectives, Policies, and Prospects*, 78–88. Westport, CT: Bergin and Garvey, 1994b.

Whitworth, Sandra. "Feminism and International Relations: Towards a Political Economy of Gender in Interstate and Non-Governmental Institutions. New York: St. Martin's Press, 1994c.

Wibben, Annick. "Feminist Politics in Feminist Security Studies." *Politics & Gender*. Vol. 7, no. 4 (2011a): 590–595.

Wibben, Annick. *Feminist Security Studies: A Narrative Approach*. London and New York: Routledge, 2011b.

Williams, Cindy. "From Conscripts to Volunteers: NATO's Transitions to All-Volunteer Forces." *Naval War College Review*. Vol. 58, no. 1 (2005): 35–62.

Wilson, Perry. "Women in Fascist Italy." In Richard Bessel (ed.), *Fascist Italy and Nazi Germany: Comparisons and Contrasts*, 78–93. Cambridge: Cambridge University Press, 1996.

Winslow Donna and Jason Dunn. "Women in the Canadian Forces: Between Legal and Social Integration." *Current Sociology*. Vol. 50, no. 5 (2002): 641–667.

Wise, James and Scott Baron. *Women at War: Iraq, Afghanistan, and Other Conflicts*. Annapolis, MD: US Naval Institute Press, 2006.

Wishnia, Judith. 1991. "Pacifism and Feminism in Historical Perspective." In Anne E. Hunter (ed.), *Genes and Gender: On Peace, War and Gender*, 84–91. New York: Feminist Press at the City University of New York.

Woodward, Rachel. "Warrior Heroes and Little Green Men: Soldiers, Military Training, and the Construction of Rural Masculinities." *Rural Sociology*, Vol. 65, no. 4 (2000): 640–657.

Wolf, Naomi. "Our Bodies, Our Souls: Rethinking Pro-Choice Rhetoric." *The New Republic*, October 16, 1995, 26–35.

Wollstonecraft, Mary. *A Vindication of the Rights of Woman: With Strictures on Political and Moral Subjects.* New York: A.J. Matsell, 1792.

Woolf, Virginia. *Three Guineas.* Orlando, FL: Harcourt, 1938.

Zalewski, Marysia. "Well, What Is the Feminist Perspective on Bosnia?" *International Affairs.* Vol. 71, no. 2 (1995): 339–356.

Zalewski, Marysia. "All These Theories yet the Bodies Keep Piling Up': Theorists, Theories, and Theorizing." In Steve Smith, Ken Booth, and *Marysia Zalewski (eds.), International Relations: Positivism and Beyond*, 340–353. Cambridge: Cambridge University Press, 1996.

Zalewski, Marysia and Jane Parpart. *The "Man" Question in International Relations.* Boulder, CO: Westview Press, 1998.

Zampini Salazar, Fanny. "Women in Modern Italy." In Mary Kavanaugh Oldham Eagle (ed.), *The Congress of Women: Held in the Woman's Building, World's Columbian Exposition, Chicago, U.S.A., 1893*, 157–164. Chicago, IL: Monarch Book Company, 1884.

Zhang, Shu Guang. *Detterence and Strategic Culture: Chinese—American Confrontations, 1945–1958.* Ithaca, NY: Cornell University Press, 1992.

Zotto, Del Augusta. "Weeping Women, Wringing Hands: How the Mainstream Media Stereotyped Women's Experiences in Kosovo." *Journal of Gender Studies.* Vol. 11, no. 2 (2002): 141–150.

Index

Page numbers in italic refer to figures and those in bold refer to tables.

Gender in a Global/Local World

Also published in this series

The Search for Lasting Peace
Critical Perspectives on Gender-Responsive Human Security
Edited by Rosalind Boyd
ISBN 978-1-4724-2096-1

Waging Gendered Wars
U.S. Military Women in Afghanistan and Iraq
Paige Whaley Eager
ISBN 978-1-4094-4846-4

Resisting Gendered Norms
Civil Society, the Juridical and Political Space in Cambodia
Mona Lilja
ISBN 978-1-4094-3431-3

No Place for a War Baby
The Global Politics of Children born of Wartime Sexual Violence
Donna Seto
ISBN 978-1-4094-4923-2

Body/State
Edited by Angus Cameron, Jen Dickinson and Nicola Smith
ISBN 978-1-4094-2449-9

Feminist (Im)Mobilities in Fortress(ing) North America
Rights, Citizenships, and Identities in Transnational Perspective
Edited by Anne Sisson Runyan, Amy Lind, Patricia McDermott and
Marianne H. Marchand
ISBN 978-1-4094-3313-2

Reshaping Gender and Class in Rural Spaces
Edited by Barbara Pini and Belinda Leach
ISBN 978-1-4094-0291-6

Federalism, Feminism and Multilevel Governance
Edited by Melissa Haussman, Marian Sawer and Jill Vickers
ISBN 978-0-7546-7717-8

Contours of Citizenship
Women, Diversity and Practices of Citizenship
Edited by Margaret Abraham, Esther Ngan-ling Chow,
Laura Maratou-Alipranti and Evangelia Tastsoglou
ISBN 978-0-7546-7779-6

Politicization of Sexual Violence
From Abolitionism to Peacekeeping
Carol Harrington
ISBN 978-0-7546-7458-0

Development in an Insecure and Gendered World
The Relevance of the Millennium Goals
Edited by Jacqueline Leckie
ISBN 978-0-7546-7691-1

Empowering Migrant Women
Why Agency and Rights are not Enough
Leah Briones
ISBN 978-0-7546-7532-7

Gendered Struggles against Globalisation in Mexico
Teresa Healy
ISBN 978-0-7546-3701-1

Encountering the Transnational
Women, Islam and the Politics of Interpretation
Meena Sharify-Funk
ISBN 978-0-7546-7123-7

The Gender Question in Globalization
Changing Perspectives and Practices
Edited by Tine Davids and Francien van Driel
ISBN 978-0-7546-3923-7 (hbk) / ISBN 978-0-7546-7322-4 (pbk)

The Price of Gender Equality
Member States and Governance in the European Union
Anna van der Vleuten
ISBN 978-0-7546-4636-5

(En)Gendering the War on Terror
War Stories and Camouflaged Politics
Edited by Krista Hunt and Kim Rygiel
ISBN 978-0-7546-4481-1 (hbk) / ISBN 978-0-7546-7323-1 (pbk)

Women, Migration and Citizenship
Making Local, National and Transnational Connections
Edited by Evangelia Tastsoglou and Alexandra Dobrowolsky
ISBN 978-0-7546-4379-1

Transnational Ruptures
Gender and Forced Migration
Catherine Nolin
ISBN 978-0-7546-3805-6

'Innocent Women and Children'
Gender, Norms and the Protection of Civilians
R. Charli Carpenter
ISBN 978-0-7546-4745-4

Turkey's Engagement with Global Women's Human Rights
Nüket Kardam
ISBN 978-0-7546-4168-1

(Un)thinking Citizenship
Feminist Debates in Contemporary South Africa
Edited by Amanda Gouws
ISBN 978-0-7546-3878-0

Vulnerable Bodies
Gender, the UN and the Global Refugee Crisis
Erin K. Baines
ISBN 978-0-7546-3734-9

Setting the Agenda for Global Peace
Conflict and Consensus Building
Anna C. Snyder
ISBN 978-0-7546-1933-8